Archaeoastronomy in
the New World

T0235737

About this book:
This volume summarises the proceedings of a conference which
took place at the University of Oxford in September 1981.
Held under the auspices of the International Astronomical
Union and the International Union for the History and
Philosophy of Science, the meeting reviewed recent progress in
the archaeoastronomy of the New World. The publisher received
the typescript for publication in January 1982.

American archaeoastronomy is growing healthily. Researchers
from different disciplines, showing an interest in native
American astronomy, have been collaborating since the early
1970s. Research paths opened by astronomers, archaeologists,
historians, anthropologists and ethnologists are converging.
In this volume a number of these paths are explored and
reviewed. The contents include a survey of progress in
understanding Maya astronomy; astronomical and calendric
practices of the Hopi and the Incas; and case studies of
Bonampak (Mexico), Chaco Canyon, and Casa Rinconada.

There is a companion volume, Archaeoastronomy in the Old
World, edited by D.C. Heggie, and also published by Cambridge
University Press.

Archaeoastronomy in the New World

AMERICAN PRIMITIVE ASTRONOMY

Proceedings of an International Conference
held at Oxford University, September 1981

Edited by
A.F. AVENI

CAMBRIDGE UNIVERSITY PRESS
Cambridge
London   New York   New Rochelle
Melbourne   Sydney

CAMBRIDGE UNIVERSITY PRESS
Cambridge, New York, Melbourne, Madrid, Cape Town, Singapore,
São Paulo, Delhi, Dubai, Tokyo

Cambridge University Press
The Edinburgh Building, Cambridge CB2 8RU, UK

Published in the United States of America by Cambridge University Press, New York

www.cambridge.org
Information on this title: www.cambridge.org/9780521125475

© Cambridge University Press 1982

First published 1982
This digitally printed version 2009

A catalogue record for this publication is available from the British Library

Library of Congress Catalogue Card Number: 82–1344

ISBN 978-0-521-24731-3 Hardback
ISBN 978-0-521-12547-5 Paperback

CONTENTS

CONTRIBUTORS

Anthony F. Aveni
Department of Physics & Astronomy
Colgate University
Hamilton, N.Y.   13346

S.C. McCluskey
Department of History
West Virginia University
Morgantown, W. Va.   26506

R. T. Zuidema
Department of Anthropology
University of Ill.   61801

G. Brotherston
Department of Literature
University of Essex
Colchester, ENGLAND   CO4 3SQ

F.G. Lounsbury
Department of Anthropology
Yale University
New Haven, CT   06520

W.B. Murray
University de Monterrey
Monterrey, NL
MEXICO

R.M. Sinclair
National Science Foundation
Washington, D.C. 20550

L.E. Doggett
U.S. Naval Observatory
Nautical Almanac Office
Washington, D.C.   20390

V.D. Chamberlain
National Air & Space Museum
Smithsonian Institution
Washington, D.C.   20560

R.A. Williamson
Office of Technology Assessment
U.S. Congress
Washington, D.C.   20510

Ms Anna Sofaer
Anasazi Project
3411 Rodman Street NW
Washington, D.C.   20008

INTRODUCTION

A. F. Aveni
Colgate University, Hamilton, New York 13346 USA

These conference proceedings reflect the healthy growth American archaeoastronomy has begun to undergo. In the early 70's, researchers from different disciplines, all sharing an interest in native American astronomical practice, gathered for the first time on the same meeting ground. They brought with them their tools, methods, and world views.

The astronomers, mathematicians and engineers sought precision both in their own work and sometimes overzealously in the ancients whom they studied. Scientific methodologies laid heavy emphasis upon carefully applying the measuring rod in the field--something which scarcely had been done before. Given an inkling from the written record that alignments did indeed exist in ceremonial space and spurred on by the conclusions of Hawkins and the Thoms at the Megalithic sites in Great Britain, scientists began to search among the ruins of North, Central, and South America to discover the extent to which Aztecs, Mayas, Incas, and Anasazi had developed an exact astronomy using architecture and the landscape to encode their observations.

Field archaeologists, who behave on the one hand like the astronomers and engineers, also reflect the approaches of the social scientist in their work. They try never to lose sight of the fact that they are dealing with real people, whose astronomy can only be said to be known and understood when its level of practice is commensurate with whatever we know from the material record of the culture.

The anthropologist and ethnologist, previously less interested and often unaware of the possible importance of astronomy in ancient culture, have turned a greater share of their attention to astronomically related myth and ritual. Interested in examining the cultural needs for different astronomies, they also have questioned their people about methods of timekeeping and the association of ritual behaviour with astronomical practice. Their queries have been rewarded, particularly among the more isolated contemporary societies. In some instances many components of the calendar have been found to survive intact into the present.

Historians and ethnohistorians have now brought their approaches to bear on the study of text, and the historical method has proven to be particularly fruitful among those civilizations where large numbers of books were written after the conquest. Many old texts, when carefully interpreted, are found to illustrate in detail the interest in sky events possessed by ancient people. Historians also have helped by examining manuscripts by ethnologists of the 19th and early 20th century who worked with native American people at a time prior to the evaporation of many ritual customs that related to the calendar.

Today, after a decade of extensive publication, critical reading, and exchange of ideas through conferences and seminars we find that many of these research paths falling within the American school of archaeoastronomy are beginning to converge. We can employ the convergence or, indeed, the divergence of arguments from astronomer, anthropologist and ethnohistorian to sort out the relative validity of various case studies. If, for example, one posits a set of astronomical alignments with high precision, good selectivity of astronomical and archaeological data, and a sound statistical approach, that argument, as positive as the theory of probability may reveal it to be, stands only on one leg. When information from another quarter suggests that astronomy was widely practiced, and identifies the events witnessed with those actually measured in the field, the case is greatly strengthened.

Of course, this scenario is a bit oversimplified. The interpretation of data takes place within every link of the chain forged by archaeoastronomical studies. Knowledge about the research in the domain of one discipline necessarily will affect that in another. If we find an abundance of orientations to a particular solstice, then we are bound to look into the texts with solstice observations in mind, and to explore the significance of such events. Of course, we must remain critical and take care not to over-interpret the data.

Such is the behaviour of interdisciplinary research and I have tried to reflect the conduct of such research over the past decade

in the Maya zone with my opening paper. There I explore the connection
between architecture and the written calendar of the Maya based on the
converging arguments that emanate from the studies of Maya epigraphy and
surveys of the ruins. While the evidence for Mars is sparse and non-
convergent, studies of the planet Venus do provide a focus. Certain
Venus events witnessed by the ancient Maya turn out to be the same as
those viewed by our ancestors, but the Maya also were interested in Venus
phenomena that had no importance in ancient Babylonia or Egypt--a fact
which reflects the relativity of cosmological viewpoints.

    In McCluskey's paper, we see historical and ethnological
studies blended to reveal the character of the precision evident in the
Hopi lunar count. Reflecting his own background as a historian of
science, McCluskey employs concepts from old European astronomy as an
analog for the Hopi case.

    Cultural anthropology represents the handle of the tool which
Zuidema wields in his study of Inca astronomy. Here the story is a very
complex one, for we know the Inca employed an unusual "writing" system
(the quipu), which we do not yet thoroughly understand. Moreover, the
mythic corpus is rather large and varied. Indeed, we find that the study
of alignments in ancient Cuzco serves as an aid to the comprehension of
the structure of a very complex society. We note that the alignment
results are not what we might have expected had we possessed only the
evidence from measurements in the field.

    One who carefully examines these contributions will learn
that primitive people often were not concerned with microscopically
precise direction setting as a means of developing accurate timekeeping.
Yet the calendar, particularly in the Maya case, is well developed,
complex, and sophisticated. We also find strong ritual constraints
placed upon the calendar keeper's activities, a fact that would escape us
totally, had we needed to rely only on the alignment data.

    Lounsbury's paper on Maya dates inscribed at Bonampak
demonstrates the tight fit between dates of actual (or ritual?) battles
and the appearance of Venus at its celestial stations. His work
demonstrates the extremes to which some ancient people could be guided by
the stars. On another front, the nature of the fitting scheme deduced by
Lounsbury provides overwhelming support for the so-called GMT Maya-
Christian calendar correlation.

Brotherston's studies regarding the commensurability of Aztec cycles follows a similar research methodology employed in Maya studies and reviewed in my paper. For the ancient Mesoamerican elite class the search for ever-larger cycles ("super numbers") into which all the smaller ones fit perfectly, seemed a consuming ambition; but, we might ask, was their approach, basically, any different from Kepler's assault on Mars which resulted in the formulation of the three empirical laws describing the motions of all the planets? Whether the form of expression be harmonic laws or super numbers, the common quest lies in the attainment of the perfect marriage between the phenomena and the mathematics.

In the contributed papers, Murray's research project on the archaic marks tallied on rock outcrops in Northwest Mexico, though in its early stages, may begin to formulate an answer to a question often raised in some of our archaeoastronomical studies: what evidence exists for the backlog of observations needed to produce primitive man's recognition of the accurate long-term cycles we read about in the codices? Evidence from an artifact of a much later period, the spiral petroglyph and associated structures at Fajada Butte, New Mexico, reported by Sofaer, offers us a possible device for recording solstitial, equinoctial, and possibly lunar dates. The petroglyph is found in the north, where we know of no written calendar at all. Clearly, we need to look for more examples of this kind of calendric practice in the petroglyphic record.

Even farther north, and east we perceive in Chamberlain's paper, which combines alignment studies and a review of old ethnological work, evidence of a knowledge of astronomy among the Pawnee as revealed in the construction of their Earthlodge. Such studies, like the more extensive work on the Gê and Warao of South America, help us to link astronomical and cosmological practices with everyday life. Finally, Williamson sums up the way in which various contributing disciplines become integrated in the Anasazi case.

The exchange of ideas between archaeoastronomers studying American and European-Megalithic material has been quite useful at this conference. Each party has given something to the other. American studies reveal the interweaving of several methodologies and the unexpected surprises that can be produced by the application of culture historical studies. Their work teaches us that good statistical results

need not necessarily guarantee the truth. Operating in an area where
they have little else to go on, the Megalithic researchers remind all of
us that our field studies must be subjected to hard statistical inquiry.
Indeed, input from non-astronomical quarters ought give us no excuse to
loosen our criteria for the selection and analysis of orientations.

      With the promise of continued cooperation between American
and European schools of archaeoastronomy pledged at the end of the
conference, through the further sharing of methods and results, and with
some effort to standardize our language and concepts, the end of the
next decade surely will leave us with a deeper understanding of ancient
cultures of the world and the role that astronomy played in them--a
common goal we all share.

      Our deepest thanks to Michael Hoskin for conceiving and
managing the conference at which these papers were presented.

      We further extend our gratitude to Queen's College, Cambridge
and to the International Astronomical Union and the International Union
for the History and Philosophy of Science for their collaborative
effort both in sponsoring the meeting and for granting funds to allow it
to take place, to Colgate Research Council for a grant of supplemental
funds for manuscript preparation, and to Lorraine Aveni for her assistance in editing and assembling the final draft manuscripts.

ARCHAEOASTRONOMY IN THE MAYA REGION: 1970-1980

A. F. Aveni
Colgate University, Hamilton, New York 13346

INTRODUCTION: DEFINITIONS AND ORIENTATION

In American studies during the 1980s, we understand archaeoastronomy to be the interdisciplinary study of ancient man's view of the
cosmos as gleaned from both the written and unwritten record. After a
stormy start beginning with the controversy over megalithic astronomy in
Great Britain fifteen years ago, this developing field has enjoyed a
decade of cooperation and progress among the contributing disciplines,
particularly in Maya studies where we have solid archaeological,
epigraphic, and ethnological evidence attesting to great mental
development.

In this paper I will review briefly recent attitudes and
discoveries about Mesoamerican (particularly Maya) archaeoastronomy,
while at the same time delineating some of the approaches that have been
used to advance our knowledge about the Maya cosmic mentality. The paper
will expand on the pre-conference position paper published in
Archaeoastronomy, No. 3 (JHA, xii (1981)). As a standard of comparison
to measure where we have gone with these studies in the 1970s I shall
refer to a paper entitled "Maya astronomy", which was read by Sir J. Eric
S. Thompson (1974) at a symposium entitled "The Place of Astronomy in the
Ancient World", jointly sponsored by the Royal Society and the British
Academy in 1972. I believe Thompson's work, more than any other,
adequately summarized the status of our beliefs about Maya astronomy a
decade ago.

Any discussion of the true meaning of Maya astronomy is
hampered by two circumstances: the deficiency of resource material and
our culture-bound orientation toward it. Most textbooks on the history of
astronomy acknowledge that the Maya were great astronomers but specific
reasons are too often omitted, the details never discussed. We always

seem to fall short of getting into the skin of the Maya astronomer. A
brief comparison of some of their astronomical views with our own reveals
part of the difficulty. We know their astronomy was time-based. Orbits,
nodes, and centrism, common astronomical concepts for us, found no place
in their cosmic scheme. Maya time was cyclic rather than linear,
heavenly periodicities being correlated with the birth, accession and
death of kings and historical epochs being marked by world creation and
destruction events; thus Cortez, the conquistador, was hailed as the
returning demi-god Quetzalcoatl-Kukulcan. Moreover, technology and
instrumentation played practically no role in the development of their
astronomy, which seems to have been largely a prognosticatory enterprise
in which religion was the driving rather than the drag force that
propelled calendric developments forward.

        Like most native American cosmologies, theirs was built
around people and their needs. Like its surviving remnant today, it was a
conservative traditional cosmology, one that sought to link every aspect
of human affairs with the course of celestial events. To abstract and
isolate natural events from the rest of human experience, to doggedly
pursue and describe each phenomenon with the utmost attainable accuracy
were goals quite repugnant to the Maya way of thought. Our difficulty is
well expressed by Thompson:

>   It is, perhaps, as irrational to expect a satisfactory
>   penetration of the mystic and emotional aura of the Maya
>   philosophy of time by a creature of twentieth-century Western
>   culture as it is to hope for a balanced, sympathetic and
>   understanding study of the ecstasy of St. Francis from the
>   pen of a militant atheist of our materialistic age. (See
>   Thompson's Introduction to Leon-Portilla (1973).)

        Given these fundamental differences between the Old World and
Native American outlooks on nature, and the fact that the written record
was almost totally demolished by the conquest and the passive survivors
engulfed by the European way of life, we can understand why Thompson made
his statement. Indeed, the atheist would have a better chance of understanding St. Francis. Yet there is hope. Like many astronomical systems
in the rest of the ancient world, control was exercised by a priestly
elite who, we can assume, desired to know with sufficient precision,
where or when natural phenomena would occur. Whether the calendric
information was used for agricultural or ritualistic purposes, whether to

regulate the ecology of an expansionist state or to order a dynastic
sequence, the grist for the calendric mill possessed its ultimate origin
in the trained eye of the observer--an eye cast upon a sky which, though
arranged quite differently, exhibited generally the same phenomena world
over. As we proceed with our review of Maya observational astronomy,
then, we shall be especially aware of two considerations: (1) that
ritualistic considerations played a powerful role in the development of
the Maya calendar (a view of pages from the Dresden Codex (1975)
(Thompson 1972) alongside a Babylonian cuneiform calendar illustrates
this principle very well, and (2) that the arrangement and course of
astronomical events was viewed in a sky tuned to tropical latitudes.

    In his paper, Thompson discussed directly our basic under-
standing in two areas relating to Maya celestial pursuits: (a) writing
and calendar (with the ever-present correlation question in tow), and
(b) alignments of buildings. Less directly, he tried to give his fellow
conferees, who were interested in the astronomy of ancient cultures of
the world, a notion of the Maya "cosmovision": the general attitude or
orientation to the astronomical realm. I hope to show that in the decade
of the '70s we have become more familiar with the Maya conception of the
heavens and that consequently we better understand the role astronomy
played in their culture; yet, as we shall see, plenty of problems remain
for the years ahead.

THE EPIGRAPHIC RECORD

    Thompson discusses the Maya calendar first, a justifiable
procedure because more attention, much of it misguided, has been focused
on the meaning of the hieroglyphs than upon any other area. The evasive
nature of the glyphs, coupled with the lack of definition of clear
problem statements (except for the correlation question), had opened the
floodgates of Maya epigraphy to all investigators regardless of
qualification: astronomers, anthropologists, computer experts, and various
other code-breakers. The procedure of decoding an artifact, made popular
in the archaeoastronomy of the early '60s (see e.g. Hawkins 1966), seemed
to condone operations on the mass of calendric data strictly from outside
the system--the way a chemist conducts a laboratory experiment. As a
result, a plethora of planetary cycles and correlations spilled out of the

pages of obscure and disparate journals which served as the only literary liaison between scholars.

Our primary resource materials for a study of Maya calendars consist principally of the fragments of four Maya codices or picture books containing hieroglyphic writing. These survived the widespread book-burnings of the conquistadors and their followers to the New World, the priests of the Roman Catholic Church who attempted to civilize the natives. The folded-screen texts once were carried from town to town by priests who made prognostications according to the apparitions of celestial bodies encoded in the data inscribed therein. Given the high percentage of space devoted to almanacs and ephemerides in these few pages of painted bark documents, we can only begin to imagine the magnitude of the lost legacy of Maya astronomy. Most of the astronomical information from the codices was rooted out of European libraries in the mid-nineteenth century, which marked the beginning of an active period in the exploration of the intellect of the Maya elite.

Other inscriptions present on the thousand or more carved monuments (mostly stelae) throughout the Yucatan peninsula provide additional information on Maya timekeeping. The pendulum of interpretation has made a full swing on the material substance emanating from this medium. Originally thought to contain only calendrical matters, these inscriptions were later believed to be more historical in nature (Proskouriakoff 1960). Only recently (see e.g. the collected papers from the Mesas Redondas de Palenque, Greene Robertson 1974, 1975, 1978, 1981) have we come to realize that a combination of human history and cosmology are imprisoned in the script.

C. P. Bowditch of the Peabody Museum, Harvard, seemed long ago convinced of the importance of astronomy in the inscriptions. He wrote in a letter to Tozzer dated 13 March 1907:

> I do want some of these men [in Tozzer's classes] to take up the study of these hieroglyphs, for I feel more and more confident, as I work at them, that they are capable of explanation. Any one who expects to bring order out of them must have a systematic course of the simpler parts of astronomy and try to tackle these Maya problems. I am sure that Goodman is wrong in thinking that the whole thing is merely a deification of numbers.

(I am indebted to Lucy B. Putnam for permission to quote the letter and to C. M. Hinsley for calling it to my attention.)

Hieroglyphic texts on pottery, many in Codex style, provide a potential fresh source for calendric material. Unfortunately, few studies have been made to date on surviving Maya ceramics, their inscriptions often being regarded as purely decorative.

Post-conquest literature consisting of indigenous sacred writings about Maya folkways, such as the Books of Chilam Balam (Roys 1946, 1949), and chronicles penned by Spanish historians living in the new world (e.g. Landa's Relación de las Cosas de Yucatan, Tozzer 1941) attest to the propagation of calendric concepts intact long after the Maya were subdued.

What did the Maya observe, how did they observe, and exactly what ends did their empiricism serve? These are the central questions for a growing number of New World archaeoastronomers.

We attack this question by analyzing one of the "super numbers" in the codices, a plan that not only sheds light on their astronomy but also enables us to demonstrate the inseparability of Maya ritual and astronomical concepts that are integrated into the calendar.

On p. 24 of the Codex Dresden (Fig. 1), we find the Long Count number 9.9.16.0.0 (transliterated). We might compare this date to a Julian Day number in our calendar. Like the odometer of an automobile, the Long Count clicks off, in a vigesimally-rooted system, day number 9.9.16.0.0 or 1,366,560 counted since the last (fourth) creation (12 August 3113 B.C., if we accept the revised Goodman-Martinez-Thompson correlation of Maya and Christian calendars).

As Lounsbury (1976, p.212) and Thompson (1950, p.226b) have shown, this number has the property that it decomposes into a relatively large cluster of low prime factors and, more central to our study, it is divisible without remainder by a collection of significant Maya calendrical numbers (italicized and identified in parentheses), some of which relate directly to observable celestial events:

Fig. 1 A page of the Venus Table in the <u>Codex Dresden</u>.
(Courtesy American Philosophical Society)

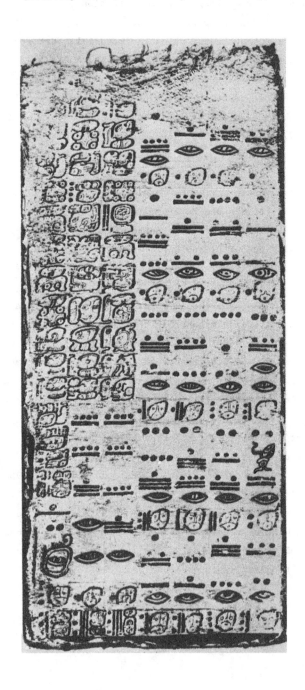

Thus $1,366,560 = 2^5 \times 3^2 \times 5 \times 13 \times 73$

$\qquad = 5256 \times 260$ (the tzolkin or sacred round)

$\qquad = 1752 \times 780 = 1752 \times 3 \times 260$ (the triple tzolkin or Mars synodic period)

$\qquad = 3744 \times 365$ (the 'vague' year)

$\qquad = 2340 \times 584 = (4 \times 584 + 4) \times 584 = 9 \times 260 \times 584$ (the Venus synodic period)

$\qquad = 2920 \times 468 = 5 \times 584 \times 468 = 8 \times 365 \times 468$ (5 Venus synodic periods or 8 vague years)

$\qquad = 18980 \times 72 = 73 \times 260 \times 72 = 52 \times 365 \times 72$ (the Calendar Round and the vague year)

$\qquad = 37960 \times 36$ (the length of the Venus Table).

Because the subject matter surrounding this super number on *Dresden* 24 constitutes a Venus ephemeris, it is no surprise to find present as multiples (a) 584^d, the Venus synodic period; (b) 2920^d, 5 Venus synodic periods and 8 vague years; and (c) 37960^d and 18980^d, the length and half-length of the Venus ephemeris on pp.24, 46-50 of the *Dresden*. As we shall see later, the number 2340 is also important in Venus computations.

There are a number of morphological characteristics about the pages of the *Dresden* table that give definite clues about precisely what aspects of this bright planet Maya astronomers actually observed: (1) Venus apparitions are broken down into four "stations"; (2) large pictures follow the tabulation of the disappearance period of the planet after the interval we call inferior conjunction; (3) the calendar functioned as a date-reaching mechanism; (4) the specific Calendar Round date associated with 9.9.16.0.0 is 1 Ahau 18 Kayab, a choice central to the proper operation of the calendar; (5) the correction scheme (p.24) rendered the calendar, otherwise astronomically incorrect over the short term, as quite precise over secular intervals.

Let us elaborate each of these facts with the goal of uncovering as much as we can about Maya observational practice.

(1) There can be no doubt (Förstemann 1901) that at the bottom of each of the five ephemeris pages are depicted the intervals:

236 ascribed to the period of Venus as morning star (263)

90 ascribed to the disappearance interval at superior
conjunction (50)

250 ascribed to the appearance interval as evening
star (263)

8 ascribed to the disappearance interval at inferior
___ conjunction (8)

584 Total

The actual mean observed intervals (subject to wide variations) are given in parentheses. A first glance at the table suggests
that the Maya were totally incapable of pinning down the great luminary.
How else could they mis-clock the length of its disappearance by 6 weeks?
But a closer look at the calendar reveals that, like the Long Count super
number, these intervals are contrived: 236 days is 8 lunar months;
250 days equals 8½ lunar months, and 90 days approximates 3 months.
Moreover, a ritual constraint in the calendar implied that only certain
named days in the 20-day cycle could be used to celebrate the transition
of the planet from one time station to the next. In fact, the named-days
for the arrival of Venus at various stations are written across the top
of the table. They are arranged in 13 lines to be read horizontally.
Having worked through modern Venus ephemerides, Gibbs (1977) has shown
that the intervals must have been chosen so as to guarantee celebration
of the appearance ritual closest to but not before first appearance as
evening or morning star. These intervals, then, link ritual dates
occurring as close as possible to the events of last and first visibility
without ever occurring during disappearance. Indeed, such a scheme
places a rather rigorous demand upon the astronomer for not only must he
observe and record celestial events, but also he must worry about
displaying his data in such a fashion that they exhibit certain commen-
surabilities with respect to non-astronomical information that emanates
from another quarter.

(2) Because the pictures always follow the 8-day disappearance in the reading of the table, they must refer to the event of the
first apparition of Venus in the morning sky (heliacal rise), which,
attended by omens written in the hieroglyphs on these pages, was the most
important event in the Venus calendar. It is stated in post-conquest
sources that the heliacal rise was, in fact, the event associated with

this particular period. (A direct reference to the 8-day disappearance
interval of Venus as Quetzalcoatl, the only interval approximating
reality, can be found in the Annales de Quauhtitlan. See Seler 1904,
pp.364-5, for a discussion.) Significantly, the Goodman-Martinez-Thompson
correlation, while derived on the basis of historical fact, equates our
super number with a date in the Christian calendar that lies within a few
days of a heliacal rise of that planet. Facts (1) and (2) taken together
suggest that we explore in detail possible means and evidence for both
observing the heliacal rise and predicting the length of the disappearance of Venus before heliacal rise.

          (3) The dynastic inscriptions at Palenque and elsewhere
recognize significant occurrences in the lives of the nobility, each
so-called "event clause" being reckoned from a starting date at the top
of a ruler's monument. Likewise the principal purpose of the calendar in
Dresden 46-50 seems to have been to record events in the Venus cycle,
though not necessarily those to which western astronomers would assign
greatest significance; i.e. it was a date-reaching mechanism. Because
the cycle-9 date also is involved with other Long Count dates reaching
into the future, the table was undoubtedly meant to predict future as
well as to record past events. Archbishop Landa tells us that certain
written documents were carried about by the priest from village to
village. These were used for the purpose of divination and
prognostication.

> They provided priests for the towns when they were needed,
> examining them in the sciences and ceremonies and committed
> to them the duties of their office, and the good example to
> people and provided them with books and sent them forth--the
> sciences which they taught were the computations of the
> years, months and days, the festivals and ceremonies, the
> administration of the sacraments, the fateful days and
> seasons, their methods of divination and their prophecies,
> events, etc. (Tozzer 1941, pp.27-8).

There is no doubt that the author and user of the calendar were not the
same person. Both Closs (1977) and Lounsbury (1978) have explicated
date-reaching schemes utilizing the information contained in the preceding
(mis-numbered) p.24, which seems to be a sort of preamble to the table,
containing user's instructions, a convenient multiplication table, and an

ingenious scheme for keeping Venus both on time with respect to the long-term calendar and on target with the correct set of ritual dates.

(4) There are three "base dates" in the Calendar Round (lowest common multiple of 260 and 365-day periods with respect to which Venus apparitions are reckoned: 1 Ahau 13 Mac, 1 Ahau 18 Kayab and 1 Ahau 3Xul. 1 Ahau is called the 'lub' of the table. It is the last named day appearing in the reading of the table and it serves as the jumping-off point from which re-entry into the beginning of the table, using a different Calendar Round base, may be accomplished. (Given the cyclic perception of time on the part of the Maya, it is not surprising to find that their tables also were recyclable.) The three base dates give us a reusable table consisting of three different ritual and vague year calendars, each entered from 1 Ahau, a Venus heliacal rise. Lounsbury (1978, p.780) has demonstrated quite convincingly that the selection of three alternative series of Calendar Round dates marking Venus phenomena resulted from small discrepancies that were discovered between the Venus canonical year (or VP) of 584 days and the true mean of the period over lengthier spans of time ($583^{d\cdot}92$). The accumulated error $0^{d}_{\cdot}08$/VP or $5^{d}_{\cdot}2$/65VP would begin to show up after one or two trips through the table (one trip = 65 VP = 104 Tropical Years). Thompson (1972) and later Closs (1977) have shown how, using the near whole multiples of the VP on *Dresden* 24, the table could be foreshortened to bring the tabulated heliacal rise more closely into line with the actual event, while still placing that event precisely on the sacrosanct 1 Ahau day. Such a process necessarily caused a shift in the vague year portion of the Calendar Round date.

Incidentally, the number 2340, one of the common multiples of our super number, also may have functioned in the correction scheme. It is just 4 × 584 + 4 days, thus allowing the user to foreshorten the table by 4 days while still relocating the same position in the sacred round (2340 is also a whole multiple of 260). Finally, 2340 is also a common multiple of distances tabulated between the three base dates (Lounsbury 1978, p.782, gives details). Thus, two basic results arise out of the considerations of facts (3) and (4): (a) the discovery of numbers like 2340 that fused ritual and astronomical observation tightly together was a clearly formulated goal of Maya calendar keepers, and (b) the determination of the time of passage of Venus from one station to the next, particularly that from obscuration by the Sun to first appearance in the pre-dawn sky, must have been a major work task for the astronomers.

(5) For want of space, statement (5) cannot be discussed in detail--the results are presented elsewhere (Thompson 1972, Closs 1977) but we must mention it because it offers some limits to accuracy that we might want to apply in our study of Maya observational methods. The upshot of Thompson's (1972) reading of the correction scheme on p.24 is the demonstrated long-term accuracy for the Venus table. Apparently, the priests deducted a total of 24 days from 301 tabulated Venus revolutions, thus placing the planet out of synchronization by 0·08 days per 481 years (in this interpretation the average synodic period is calculated at $583^{d}.92026$). Overall, in spite of its short-term inaccuracy owing to compromise in the face of sacred dictates, the long-term accuracy of this warning table is geared to anticipate heliacal risings of Venus, with early predictions being favoured over late ones.

ALIGNMENT STUDIES AND OTHER EVIDENCE BEARING ON THE INFORMATION CONTAINED IN THE WRITTEN CALENDAR

It is interesting that in his 1972 paper Thompson discusses this subject under the title "alinements [sic] and measures" (p.94) and allots only 2½ of his 15 pages to it. His statement that the "Divergence of one wall from another presumably resulted from the inability of the Maya to lay out true right angles" demonstrates the generally low regard in which the astronomical orientation hypothesis was held in the early 70s. Thompson's indictment of celestial orientations is especially surprising when in the very text of his own essay he tells us about Motolinia's statement (1971, Pt. I, ch. 16, para. 89) proving the equinoctial orientation of the Templo Mayor of Tenochtitlan, and Lincoln's (1942) investigation of the modern Ixil Maya double astronomical alignment at Nebaj. He even quotes detailed measurements of structure alignments at Tikal made by Tozzer, who must have had some thoughts about building orientation to have recorded his measurements so precisely. But all is dismissed with the statement "One suspects that...variation arises from sloppiness, for such variability occurs all over the Maya area and is a warning against crediting the Maya with intention and precision for every significant orientation noted".

Recent evidence (e.g. Hartung 1977) suggests very definite
intention and even remarkable precision in a few cases; on the other hand
the over-emphasis upon "scientific" precision may be a detriment to our
understanding of the Maya "cosmovision". After all, precision has been
the hallmark of excellence by which to judge most ideas proposed by the
investigators of the astronomy of prehistoric Britain. But in the
Americas we have more than our own instruments and standing stones to
work with. The ceremonial centre was the sacred turf of the Maya priest.
As the embodiment of heaven on earth it should incorporate divine
celestial principles. I see a likeness between the deviations from
symmetry in the Maya architecture and Lounsbury's contrived numbers that
appear to deviate from perfect astronomical values for ritualistic
reasons. I believe the architectural astrogeometry likely served as
derivative of the information that went into the making of the calendar.
The asymmetries of the Caracol of Chichén Itzá, of Building J at Monte
Alban, the skew of the base of the Governor's Palace at Uxmal, all
emphasize *not scientific precision in the use of astronomical instruments*
(an applaudable western technological feat), but a total awareness of the
natural environment and a desire on the part of the Maya to embody
nature's course in the earthly works.

Let me summarize the new evidence on orientations in a few
well-documented cases referring specifically to the Venus calendar.
In each case we shall see how the archaeological record and astronomical
techniques together are being utilized to bring forth the evidence.

Returning to the results of the previous section, we ask:
what observations are necessary to predict the heliacal rise of Venus and
is there any evidence in the surviving record bearing on such
observations? Unlike the Babylonian case there are no remnant "notebooks"
delineating a record of sightings. Indeed, the written record both on
manuscript and in stone is enveloped entirely in the complex system of
interrelated Maya time cycles. It is refined data. Nowhere among the
dots, bars and hieroglyphs do we find a direct reference to a single
sighting.

An important clue to the question lay hidden for a long time
in the standing architecture in the form of alignments toward astronomical events transpiring at the horizon. It is perhaps more logical to
anticipate horizon observations in the astronomies of tropical latitudes.

Diurnal motion in the higher latitudes may be characterized as poloidal, but objects in tropical skies rise and set along nearly vertical tracks with respect to the local horizon. Consequently, stars keep relatively fixed azimuths as they execute daily rise and set motion. Near the equator, such directions can be established as celestial reference points much more easily than would be the case at temperate latitudes. The zenith-nadir axis and its fundamental reference circle, the horizon, provide a natural reference frame. Can this be the reason why the equatorial reference frame or the zodiac are not as well developed in the Americas as in China or the Near East where they flourished?

As I have shown (Aveni 1980), the precise observation of the time and place of the last disappearance of Venus over the western horizon constitutes an index of the length of time about inferior conjunction that the planet remains obscured by the Sun. Thus it is logical to anticipate that a Venus watcher in the tropics might set up horizon alignments to the west. A second horizon observation might pertain to the standstill positions or horizon extremes for the planet. While extremes occur on an annual basis because Venus more or less follows the Sun, a seasonal cycle of *great* extremes, nearly as large in horizontal extent as that of the Moon, is repeated almost exactly after an 8-year interval, the same period tabulated in *Dresden* 46-50.

Could the Maya penchant for constructing buildings with asymmetric ground plans be attributed to the need to orient toward horizon phenomena?

Indeed, the possibility for precisely astronomically oriented buildings was taken quite seriously by Morley as early as 1909:

> Though I only worked one night at Kabah, I found out that the building in which we camped [Casa No. 1, Stephens] also ran east of north as did all the other houses examined at Chichén Itzá and Uxmal. Moreover, judging by eye from this building and as far as I could make out all the other buildings at Kabah, twenty in number at least, also are now east of north in their orientation. Should this be found true at other sites in Yucatan I believe we are near to establishing the point that a North Star orientation was practiced by the ancient priest-rulers.

(Letter to Bowditch, 3 May 1909. I thank the Peabody Museum for
permission to quote and C. M. Hinsley for calling this letter to my
attention.) It is a pity that such notions about the alignment of
buildings were not followed up in the explosive developmental period of
Maya archaeology around the turn of the century.

Fig. 2   The Palace of the Governor at Uxmal
(Photo courtesy H. Hartung)

Our transit measurements of ancient Yucatecan buildings show
that while many adjoining walls are non-rectangular, the Maya could do a
decent job of making right-angles and of erecting perfectly parallel
facades when they needed to. One instance where astronomical necessity
presented itself is the Palace of the Governor at Uxmal. Though the
short side walls are out of line, the long front and rear walls and the
principal doorway face 28°05' S of E, which lies within 2 minutes of arc
of the southerly great extreme position of Venus rising in A.D. 750.
(Curiously, the great palace at Santa Rosa Xtampak, 70 km to the south,
possesses exactly the same orientation. A second Venus orientation also
occurs at Uxmal, toward southerly standstill on the western horizon. It
proceeds from the western doorway of the House of the Dwarf to the centre

of the ballcourt to the center of the principal doorway of the West
group.) The entire Palace, having been built on an artificial mound of
400 square metres extent, is skewed by 20° in the clockwise sense with
respect to the other buildings at Uxmal. The principal pyramid at the
ruins of Nohpat, a structure of 25 m elevation, bears 28°13' S of E from
the central doorway of the Palace at a range of 6 km so that it
effectively marked the key "turn-around" point in the Venus calendric
cycle in the distant landscape about the time the Governor's Palace was
built. Venus iconography adorning the facade of the Governor's Palace
provides added support for the Venus alignment hypothesis (Seler 1917,
p.135). The cheeks of the Chac (rain-god) masks above the cornice are
adorned with Venus glyphs (T510), the same ones appearing on pp.46-50 of
the *Dresden*. (For a discussion of Venus-rain references see Closs *et al.*,
in press.) Nearly four hundred such symbols in all can still be counted
today. Moreover, the mask at the NE corner of the building displays an
8-symbol in dot-bar notation below the eye, which Seler had already
interpreted as the 8-day disappearance interval nearly half a century
before our orientation studies (see Figure 3).

Fig. 3   Carved number 8--3 dots suspended from a stylized bar
over the eye of the Chac mask on the SE corner of the
Palace of the Governor
(Photo courtesy of H. Hartung)

How did the builders of the Governor's Palace determine the
line of the facade so that the perpendicular to the central doorway would
align exactly with Nohpat and the southerly extreme rising point of the
planet Venus? It seems unlikely that the Maya architects accomplished
this task by sighting along the short side walls of the building and then

constructing the long walls perpendicular to them. The short walls are simply not parallel. After the builders erected the large artificial platform, they probably laid out a sight line to Venus at its southerly extreme, perhaps using a series of sticks positioned by visual sightings of the planet at horizon as viewed from the intended location of the doorway of the Palace. Once the position of the central doorway of the Governor's Palace was incorporated into the plan it could be employed to determine precisely where to locate the Nohpat pyramid. Later, the portion of the baseline nearer Uxmal could be fixed with a cord, and then a perpendicular cord laid down to mark the alignment of the eastern (front) facade that was intended to overlook Nohpat. Finally, a parallel cord could be arranged to delimit the western (rear) facade of the Governor's Palace. (The long front and rear facades are antiparallel by only two arc minutes.) In the absence of additional evidence about Maya construction technology and planning, it seems dangerous to push this already speculative scenario much further at this time.

Fig. 4   The Nunnery at Uxmal
(Photo Courtesy of H. Hartung)

Still more Venusian information can be found in the iconography of another Uxmal building. Lamb's (1980) count of the mosaic facade on the East, West, and North buildings of the Nunnery of

Uxmal (Fig. 4) reveal the same philosophy of commensuration delineated in the codices. He found that the X's in the bounded double-headed serpent and bars on the eastern facade of the Nunnery add up to 584. Furthermore, Maya astronomers may have employed other elements of the Nunnery facade to tally the vague year, lunar synodic and sidereal months, possibly even the Venus sidereal period.

The Nunnery complex of Uxmal serves as a useful example to demonstrate that Maya builders also were concerned with geometrical precision. (See Hartung 1972 and Aveni & Hartung (in press) for a full discussion of this problem.)

Fig. 5 Plan and section of the Nunnery at Uxmal showing geometrical alignments (Diagram by H. Hartung)

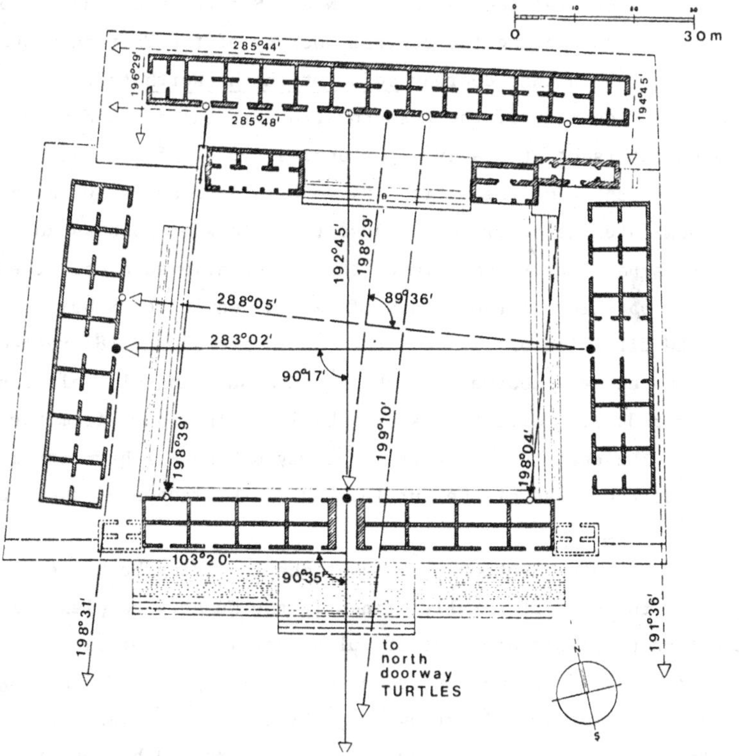

Located in the NE section of Uxmal adjacent to the Pyramid of the Magician, the Nunnery is comprised of four rectangular buildings that enclose an irregular courtyard approximately 60^m by 80^m. But a look at the plan (Fig. 5) showing the relative placement of the 4 component structures reveals that sloppiness, even disregard of detail, hardly could have resulted in the asymmetry and non-rectangularity at the junctures of the 4 major buildings that we observe today (the angles are 84°49' SW, 92°43' NW, 94°12' NE and 88°16' SE). Indeed, the 4 buildings appear so out of line that they force us once again to entertain the notion that misalignment was deliberate and planned.

Consider the following axes in Fig. 5:

a) that formed by a pair of lines, one connecting the central doorways of the E and W structures, the other the central doorway (the archway) of the S structure and that doorway adjacent on the West to the central doorway of the N structure, (solid lines in figure), and

b) that formed by another pair of lines, one connecting the central doorway of the N and S structures and the other connecting the central doorway of the Eastern structure with that doorway adjacent on the North to the central doorway of the Western structure (dotted lines in figure).

These axes intersect at nearly right angles (90°17' for a) and 89°36' for b)) near the center of the patio. Moreover, the S and W buildings are made to dominate the other two because their facades possess an orientation that is parallel to one arm of the aforementioned axes. The W building is parallel to the NS arm of axis b) while the S building aligns with the EW direction of axis a) (the errors are 2' and 18' respectively). Finally, the extreme doorways on the N and S buildings lie along parallel lines (error 35'), a direction shared by 1) a line taken from the doorway adjacent on the east to the Central doorway of the Northern structure prolonged to the House of the Turtles and 2) the facade of the House of the Turtles itself.

The aforementioned observations involving both the double set of axes and the sets of parallel lines in the Nunnery complex illustrate a construction principle or a set of rules that is both precise and consistent, even if somewhat strange to us. The precise right angles we sought at the corners or junctions of buildings now appear in an abstract way in the open space, at the center of the courtyard between the buildings. Similar interbuilding relationships that may have been governed by geometrical considerations are found to occur at Tikal, Copán, and Chichén Itzá (Hartung 1972, 1977). Of course, the process of discovery of the pair of right angles as we relate it can be very different from the manner in which the Maya conceived and planned the Nunnery. Like all works of art, we see and interpret it through our own eyes.

At Chichén Itzá the Maya-Toltecs preserved Venus orientations in the Caracol, a round building dedicated to Venus in the form of the wind god. The structure is perched on a two-tiered quadrangular base of both odd shape and orientation (Fig. 6).

Fig. 6   The Caracol of Chichén Itzá
(Photo courtesy of H. Hartung)

But why do we select Venus instead of the more obvious sun or moon as the best match for the building orientations? In Table 1 we display the azimuths of the 4 orientations in question together with the azimuth difference between each orientation and 3 horizon points: the summer solstice sunset, the lunar extreme (+ε + ɩ) set position and the Venus great extreme set position in the period AD 900-1000, the time the building was erected. Clearly the Venus alignment gives the best fit.

Table 1.   Probable Venus Alignments in the Caracol of
Chichén Itźa

Building Alignment	Azimuth of Building Orientation	Azimuth Difference Between Building Orientation And Setting Extreme* of		
		Sun	Moon	Venus
Perpendicular to base of Lower Platform	297°24'±6'	+1°44'	-3°10'	-1°13'
Perpendicular to base of Stylobate Platform inset into Upper Platform	298°00'±30'	+2°20'	-2°34'	-0°37'
Window 1 inside left to outside right jamb	298°53'±10'	+3°13'	-1°41'	+0°16'
Window 2 inside left to outside right jamb	297°49'±10'	+2°09'	-2°45'	-0°48'

*at last gleam

But there are more compelling reasons for our choice than the
analysis of alignment data. First, we know that round temples were
erected for the purpose of venerating one form of Quetzalcoatl-Kukulcan,
who is the same god identified with Venus in the Maya inscriptions.
Moreover, he is the same god pictured on every page of the Venus table in
the Dresden Codex, a Venus ephemeris incorporating 8-year cycles of the
motion of the planet. The Dresden table, which, we will recall from our
earlier discussion, highlights the predicted apparition of Venus as
morning star in the east after it becomes lost in the solar glare, is
known to belong approximately to the same provenience both in space and
time as the Caracol. Finally, the Maya would have had a compelling
functional reason to sight Venus disappearing in the west if they wished
to devise a Dresden-type calendar for the time and place of Venus'
disappearance provides information about where and when it will reappear
in the east. These converging interpretations, derived from the examination of the architecture, the inscriptions, the historical record and
the motion of Venus lead us to suggest that the Caracol may have been the
very observatory they employed to derive the data we find written in the
Maya Dresden table.

Fig. 7 Sculpted doorway of the Temple of Venus,
Temple 22 at Copan
(Photo courtesy H. Hartung)

Fig. 8 Enlarged portion of upper left portion of sculpted
doorway showing the same Venus symbol that appears
in the Venus Table of the Dresden Codex
(Photo courtesy of H. Hartung)

Finally, Temple 22 at Copan has long been called the Temple of Venus because of the dominating T510 glyphs appearing in high relief on its sculpted doorway (see upper left in Fig. 7 and the enlargement in Fig. 8). It possesses a single narrow vertical window on the west front (Fig. 9). This window has yielded an intriguing set of Venus alignments of possible agricultural importance (Closs *et al.*, in press). We discovered that all the great northerly extremes of Venus occurring at 8-year intervals between AD 700 and AD 900, when this section of the building was erected, were visible in April or May. This is the period which was singled out as particularly appropriate for celebration in honor of the Holy Cross in the Chan Kom region. More specifically, within this period the Venus extremes tended to occur in late April or early May. This coincides with the usual onset of the rainy season in the Chorti area. In fact, the Venus extremes cluster around the 8-day interval between 25 April and 3 May, the formal ceremonial period for the coming of the rains and the planting of the new maize. Indeed, between AD 700 and 781, 8 of the 10 extremes of one set of 8-year Venus cycles lay within this interval. Remaining extremes were found to be within 3 days and 1 day of it. Between AD 824 and AD 897, 7 of 10 extremes of another set of 8-year cycles lay within the interval, the others lying within 6, 2, and 2 days of it. In the years between AD 781 and AD 824, 5 extremes of the first set and 5 of the second set lay an average of 7 days from the 9-day interval between 25 April and 3 May. Thus we have affirmed that great northerly extremes are seasonal phenomena occurring at the time associated with the beginning of the rainy season in Copan.

We have focused upon Venus in the written and unwritten record in order to provide an intimate understanding of the relationship between "theory" and observation for one particular set of phenomena. The examples illustrated suggest that architecture was for the Maya what pen and parchment were for the Greeks, a medium of expression of geometrical knowledge. For completeness, we must summarize, if only briefly, what other astronomical phenomena attracted the Maya and how the written knowledge of these phenomena is grounded in observational evidence. We can be sure the Maya were interested in eclipses. The table on *Dresden* 51-58 leaves no doubt the Maya possessed the capacity to provide a reliable warning system for lunar eclipses and surely for solar eclipses occurring at the same solar node passage. The scheme, for which all the details have yet to be spelled out even after a century of study, is based upon an arrangement, familiar to students of Babylonian astronomy, of lunar synodic months taken in groups of 6, with occasional strategic insertion of groups of 5 into the count. As we have already stated, the Maya had no need of the orbital concepts of node or ecliptic limit nor do we have any evidence that they used the lunar horizon extrema that surface so often in the literature on megalithic astronomy. Rather, New World astronomers simply seem to have noted the time occurrence of New (or Full?) Moon. The lunar eclipse table which, incidentally, contains the saros as a hidden factor, yields an averaged lunar synodic month of $23\overset{d}{.}530864$ (off 7 seconds per month); 405 moons were tallied over a period of 11,958 days, 2 days short of what a modern astronomer would tabulate. Enough solar horizon references are present in alignment studies (Aveni 1980, ch. V, summarizes them) and tropical year counts in the inscriptions (Kelley and Kerr 1974, Teeple 1930) to establish that an accurate year of the seasons was reckoned. (Long-term accuracy averaging a few minutes per year is claimed to have been attained (Teeple 1930, p.74). While there is no strong evidence that intercalation was practiced, Kelley has

Fig. 9   Plan of Temple 22, the Temple of Venus at Copan, showing its Venus facing window (arrow) (Plan courtesy H. Hartung after an earlier map by A. Trik)

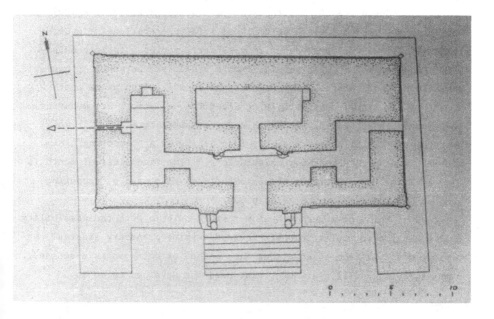

recorded quarter-year intervals between Calendar Round dates on stelae. Maya astronomers seem to have exhibited some concern over the observation of the tropical year anniversary within the vague year calendar, but at the surface they seem to have been quite less preoccupied than their western counterparts with establishing a calendar that ran precisely with the seasons.

Once the Sun, Moon and Venus have been discussed, we pass into a far less certain realm regarding concrete data. Willson(1924) called attention on pp.43b-45b of *Dresden* to a possible Mars table containing a "Zip monster" with upturned snout. But as Thompson (1972, p.108) pointed out, the format of these pages is so substantially different from that of either the lunar or Venus tables as to leave some doubt, *e.g.* there would have to be 10 recognizable stations of the planet if it is a real Mars table. Furthermore, the near coincidence of the mean Mars synodic interval (779^d.94) with the triple tzolkin (780^d), recorded so prominently in this table, has been used to argue the position that only a simple whole multiple of this sacred cycle actually was being tabulated. In the light of our earlier discussion of the contrived nature of the super number of the codex, today one might prefer to take the position that the Mars period is exactly the kind of cycle the Maya astronomer would have looked for. Surely this table deserves further study.

Could the 10 hypothetical Mars stations (10 × 78 = 780) tabulated on *Dresden* 43b-45b be zodiacal reference stations? In *Codex Paris* (1968), 23-24, we have a firm case for a zodiacal table consisting of 13 animal figures pendant from a banded sky-serpent. Linguistic and ethnohistoric sources give us star and constellation names that correspond to astral referents lying along the ecliptic. Accompanying these pages is an 1820-day lunar calendar arranged in a 5 × 13 × 28^d pattern, the latter number possibly functioning as a "computing" sidereal month convenient for calculations. (The true sidereal month is 27.3216 days long. There is evidence the Maya also used a "computing year" of 364 days (Thompson 1972, p.109).)

While no modern astronomer would quibble with the possibility that non-heliocentric astronomers could determine planetary sidereal periods directly from observations (one method of doing so is discussed in Aveni (1980, ch. III), we wonder what need the Maya had to do so.

Possibly the assignation of various planet-star conjunctions was ritually intermingled with the course of dynastic history. Like all the other observations discussed in this paper, these sidereal cycles are also well within the reach of naked-eye technology. Some investigators have expressed surprise that the other planets are absent from the records. I think it safer to assume that, though the record is sparse, the evidence lies before us, hidden in the tangled morass of ritual, the mastery of which has impeded us for so long. To date the evidence is not over-powering. (Kelley (1980) brings forth some new arguments on the Mercury synodic period in the codices.) Yet, the Maya were too perspicacious to pass up the kind of challenge Mars gave Kepler. Neither could bright Jupiter nor steady Saturn have escaped them. Perhaps the post-conquest literature of Yucatan and the contemporary ethnographic record ought to be brought to bear in archaeoastronomical investigations that might be conducted in the future. The Books of Chilam Balam are rarely related to a discussion of the astronomical content of the codices and aside from the brief survey of Cakhiquel Maya astronomical habits by Remington (1977), the recent studies of the use of the 260-day calendar among the Quiche by Tedlock (n.d.) and the work of Vogt (1969) and Gossen (1974), I know of few other efforts to look into the cosmological remnants of Maya astronomy. Perhaps the approach of Zuidema (1977) and his students (e.g. Urton 1981) in the Andean field can serve as an example to follow. As an incentive, I suggest Gossen's (1974) and Marshack's (1974) studies of the Chamula calendar board, which revealed the conservative nature of contemporary calendar keepers, some of whom still tenaciously adhere to Pre-Columbian principles of recording and measuring time. And finally, as a challenge, the recent *Current Anthropology* paper by Turton and Ruggles (1978) demonstrates the vast complexities, many of them non-astronomical, which serve as obstacles to the comprehension of time-keeping systems among indigenous non-western cultures.

CONCLUSIONS AND CONTRASTS

It might be instructive to contrast in a general way Maya astronomical methods with those of their ancient Old World predecessors, the Babylonians, about whom Aaboe (1974) and his colleagues have written in such eloquent detail for so long a time. Though they did not employ saw functions and synodic arcs in a space-based system, like the Babylonians, the Maya created astronomical tables that served principally to predict events, not to calculate durations. This similarity ought not surprise us, for the goal of any functional calendar is to plot the course of natural events and to make possible the accurate prediction of when they should arrive. In most societies it is the repeated recurrence of the celestial cycle, not the length of time between successive events, that first captures the interest.

Also like the Babylonian tables, the Venus and lunar eclipse ephemerides entailed calculations employed for warnings, if not for exact predictions. They were definitely not a record of observations (Aaboe 1974, pp.27,30). Even the scheme for the attainment of long-term precision is similar (Aaboe, p.35).

On the other hand there are some significant differences between Babylonian and Maya astronomy. The latter, like most tropical astronomies, was based almost exclusively on observations of events at horizon, while the former seemed to have been centered on the ecliptic, like many of the indigenous astronomies of the higher latitudes (Aveni 1981). Retrograde motion and conjunctions, typical planet-to-star or planet-to-Sun references in the astronomy of the classical world are submerged in Maya observational astronomy. These concepts are replaced by a set of planetary stations pivoted about the appearance of planet at horizon, a natural development that might be anticipated in the celestial environment of the tropics. Also, Maya calendars are rife with non-astronomical constructs. While the classical Old World notion of isolating astronomical phenomena from the rest of the natural world provided certain advantages (*e.g.* the discovery of the saros and Metonic cycles), the Maya scheme of integrating natural phenomena in the biosphere with astronomical phenomena on the celestial sphere also had its positive features, namely a more unified and harmonious view of the natural world (Leon-Portilla 1973).

Finally, what about model building, the interplay between theory and observation and the attainment of a 'true' scientific astronomy? Having grappled with this question for a long time, I am forced to take the view that, whether one uses Aaboe's multi-level definition of scientific astronomy or McCluskey's (n.d.) more liberal criteria for categorizing this sort of intellectual endeavour, all of these notions are so purely western that they should be applied only with the greatest caution. The old literature shows all too many attempts to frame Maya astronomy within a western mental border; the results have

been disastrous, as Thompson (1974) has shown. I cannot believe that if
the Maya were left to themselves they would have developed the calculus,
heliocentric astronomy, or our version of scientific explanation. In
McCluskey's recent paper, he has attempted to separate scientific aspects
of Hopi astronomy from the religious or magical residue. He does so
because he seeks a vehicle for understanding *their* astronomy in *our* terms.
McCluskey's attempt is an admirable one, but if it is carried to extremes
the results may be that we shall kill the goose that laid the golden
eggs. For Maya ritual and astronomy, like observer and experiment in
quantum physics, are inextricably intertwined. It remains questionable
to what extent we can expect to derive significant information about
complex non-western systems of astronomy by focusing only upon the
observational aspects of those systems. We must strive to comprehend the
whole picture as seen through their eyes. Anthropological studies must
be a part of the picture. At least this seems to be the moral of any
in-depth inquiry into the story of Maya archaeoastronomy.

   Our notion of scientific explanation in astronomy has
followed a long and complex evolutionary path. It is founded on the
premise that one can totally divorce astronomical phenomena not only from
social and religious contexts, but also from events transpiring in other
realms of the natural world. We have become abstract, reductionist
model builders and there are certain rules by which we play our game of
science. While there is some evidence that the Maya acquired and
manipulated certain cognitive impressions the way we did, there is no
indication that they took further steps along the circuitous evolutionary
path we ended up following. Yet they remain worthy of our attention. Of
the network of complex roadways leading from the eye to the mind, how
could we really expect the astronomies of two cultures so diverse and
forever isolated from one another to have pursued the same course?

   I began this overview with the wisdom of Thompson and so
shall I end it:

   Had astronomers interested in Maya astronomy such a background [*i.e.*, Maya culture, history, archaeology], we would
   be spared assertions that the Maya calendar was in full swing
   before 3500 BC, when in fact, agriculture in the New World
   had hardly got under way, and the Maya were still over 3000
   years short of developing an identity. We would also be

saved from consequent wild deductions that because the Pleiades were on the celestial equator at that date, the Maya were then probably living in or near Peru (Smiley 1960). Needless to say, there is not the slightest evidence that the Maya or anyone else in Middle America had then any sort of reliable time reckoning, or that the ancestors of the Maya were then living south of the equator. In the light of such deductions, which one can only designate as highly intemperate, and of other material discussed above, one is inclined to say that Maya astronomy is too important to be left to the astronomers.

REFERENCES

Aaboe, A. (1974). Scientific astronomy in antiquity, Philosophical Transactions of the Royal Society of London, A cclxxvi, 21–42.
Aveni, A.F. (1980). Skywatchers of Ancient Mexico. Austin and London: University of Texas Press.
Aveni, A.F. (1981). Tropical archaeoastronomy, Science, 213, 161–171.
Aveni, A.F. & Hartung, H. (in press). Precision on the layout of Maya architecture, Proc. N.Y. Acad. Sci.
Aveni, A.F., Gibbs, S.L. & Hartung, H. (1975). The Caracol Tower at Chichen Itza: An ancient astronomical observatory?, Science, clxxxviii, 977–85.
Closs, M.P. (1977). The date-reaching mechanism in the Venus table of the Dresden Codex , In Native American Astronomy, ed. by A.F. Aveni, pp. 89–99, Austin and London: University of Texas Press.
Closs, M.P., A.F. Aveni and B. Crowley (in press). The planet Venus and Temple 22 at Copan, Indiana.
Codex Parisianus (Codex Paris) (1968). Akad. Druk U. Verlag. Graz.
Codex Dresdensis (Codex Dresden) (1975). Akad. Druk. U. Verlag. Graz.
Förstemann, E.W. (1901). "Commentar zu Mayahandschrift der Koniglichen offentlichen Bibliothek zur Dresden", trans. by Selma Wesselhoft and A.M. Parker and published in 1906 as "Commentary on the Maya manuscript in the Royal Public Library of Dresden", Papers (of the Peabody Museum of American Archaeology and Ethnology), iv.
Gibbs, S.L. (1977). Mesoamerican calendrics as evidence of astronomical activity, In Native American Astronomy, ed. by A.F. Aveni, pp. 21–35, Austin and London: University of Texas Press.
Gossen, G.H. (1974). A Chamula solar calendar board from Chiapas, Mexico, In Mesoamerican Archaeology: New Approaches, ed. by N. Hammond, pp. 217–53, London: Duckworth.
Greene Robertson, M. (1974). Premira Mesas Redonda de Palenque, vols. 1,2 : A Conference on the Art, Iconography and Dynastic History of Palenque. Pebble Beach Ca: R.L. Stevenson School.
Greene Robertson, M. (1975). Segunda Mesas Redonda de Palenque, vol. 3. Pebble Beach Ca: R.L. Stevenson School.
Greene Robertson, M. (1978) Tercera Mesas Redonda de Palenque, vol. 4. Monterey Ca: Herald.
Greene Robertson, M. (1981). Cuarta Mesas Redonda de Palenque, vol. 5. Austin: University of Texas Press.
Hartung, H. (1972). Consideraciones sobre los Trazos de Centros Ceremoniales Mayas, XXXXVIII Acts of the Int'l Cong. of Americanists, vol. 4: 14–26. Kommissionsverlag K. Renner, Stuttgart.
Hartung, H. (1977). Ancient Maya architecture and city planning, In Native American Astronomy, ed. by A.F. Aveni, pp. 111–30, Austin and London: University of Texas Press.
Hawkins, G. (1966). Stonehenge Decoded, New York: Delta Dell.
Kelley, D.H. (1980). Astronomical identities of Mesoamerican gods, Archaeoastronomy, no. 2 (Journal for the history of astronomy, xi), S1–S54.
Kelley, D.H. and K. Ann Kerr (1974). Mayan astronomy and astronomical glyphs, In Mesoamerican Writing Systems, ed. by E.P. Benson, pp. 179–215, Washington, D.C.: Dumbarton Oaks, Trustees for Harvard University.
Lamb W. (1980). The Sun, Moon and Venus at Uxmal, American Antiquity, xlv, 79–86.
Leon-Portilla, Miguel (1973). Time and reality in the thought of the Maya, trans. by Charles L. Boiles and Fernado Horcasitas, Boston: Beacon Press.
Lincoln, J.S. (1942). The Maya calendar of the Ixil of Guatemala, Carnegie Institution of Washington Contributions to American anthropology and history, vii, 97–128.
Lounsbury, F. (1976). A rationale for the initial date of the Temple of the Cross at Palenque, in Greene Robertson, 1975, vol.3, 211–24.
Lounsbury, F. (1978). Maya numeration, computation and calendrical astronomy. Dictionary of Scientific Biography, vol. 15, suppl. 1, ed. by C.C. Gillispie, pp. 759–818. New York: Scribner's.
Marshack, A. (1974). The Chamula calendar board: An internal and comparative analysis, In Mesoamerican Archaeology: New Approaches, ed. by Norman Hammond, pp. 255–270, London: Duckworth.
McCluskey, S. (n.d). Hopi astronomy and the nature of traditional science, paper read at Atlantic Regional Meeting of the History of Science Society, University of Maryland, April 1980.

Motolinia, Fray T. de B. (1971). Memoriales o Libro de las cosas de Nueva Espana y de los Naturales de Ella, ed. by E. O'Gorman, Mexico City: Universidad Nacional Autonoma de Mexico.

Proskouriakoff, T. (1960). Historical implications of a pattern of dates at Piedras Negras, American Antiquity, xxv, 454–75.

Remington, J.A. (1977). Current astronomical practices among the Maya, In Native American Astronomy, ed. by A.F. Aveni, pp. 75–88; Austin and London: University of Texas Press.

Roys, Ralph L. (1946). The book of Chilam Balam of Ixil, Carnegie Institution of Washington, Notes on Middle American Archaeology and Ethnology, no. 75, 90–103.

Roys, Ralph L. (1949). The prophecies for the Maya tuns of years in the Books of Chilam Balam of Tizimin and Mani, Carnegie Institution of Washington Contributions to American Anthropology and History, x, 153–86.

Seler, E. (1904). Venus period in the picture writings of the Borgian Codex group, Bureau of American Ethnology, Washington Bulletin, xxviii, 355–91.

Seler, E. (1917). Die Ruinen von Uxmal, Abhandlungen der Koniglich Preussischen Akademie der Wissenschaften, Berlin, iii.

Smiley, C.H. (1960). The antiquity and precision of Mayan astronomy, Royal Astronomical Society of Canada Journal, liv, 222–76.

Tedlock, B. (n.d.). Sound texture and metaphor in Quiche Maya ritual language, paper read at XLIII International Congress of Americanists, Vancouver, B.C., August 1979.

Teeple, J.E. (1930). Maya astronomy, Carnegie Institution of Washington Contributions to American Archaeology, i, no. 2, 29–115.

Thompson, J. Eric S. (1950). Maya Hieroglyphic Writing, Norman: University of Oklahoma Press.

Thompson, J. Eric S. (1972). A commentary on the Dresden Codex (American Philosophical Society, Philadelphia, Memoirs, no. 93).

Thompson, J. Eric S. (1974). Maya astronomy, Philosophical transactions of the Royal Society of London, A cclxxvi, 83–98.

Tozzer, A. (1941). Landa's Relacion de las Cosas de Yucatan, Papers, xviii, Cambridge, MA: Peabody Museum, Harvard University.

Turton, D. and C. Ruggles (1978). Agreeing to disagree: The measurement of duration in a southwestern Ethiopian community, Current Anthropology, xix, 585–600.

Urton, G. (1981). At the Crossroads of the Earth and Sky, Austin and London: University of Texas Press.

Vogt, E.Z. (1969). Zinacantan: A Maya Community in the Highlands of Chiapas, Cambridge, MA: Belknap Press of Harvard University.

Willson, Robert W. (1924). Astronomical notes on the Maya codices, Papers,vi, no.3, Cambridge, MA: Peabody Museum, Harvard University.

Zuidema, R.T. (1977). The Inca calendar, In Native American Astronomy, pp. 219–59, ed. by A.F. Aveni, Austin and London: University of Texas Press.

HISTORICAL ARCHAEOASTRONOMY: THE HOPI EXAMPLE

Stephen C. McCluskey
West Virginia University, Morgantown, West Virginia   26506
U.S.A.

Like many of us, I first became interested in early astronomies through the works of Sir Norman Lockyer (1906), which I first
encountered in a course in the history of ancient science. I then moved
on to Gerald Hawkins' more recent account of Stonehenge (Hawkins & White,
1965) and the meticulous surveys of Alexander Thom (1967, 1971, Thom &
Thom, 1978) and his followers. But as a historian, I was immediately
troubled by the long step taken in these discussions from archaeological
evidence to astronomical inference and began to look for historical
evidence that would document more firmly the existence, nature, and
precision of astronomies of this type.

A series of accidents led me to consider the Hopi of Northern
Arizona, who remained isolated from Euro-American contact during the
Spanish period--at least in comparison to the Christianized Pueblos of New
Mexico. Early in the period of contact with Anglo-Americans a series of
careful observers including Jeremiah Sullivan, Alexander M. Stephen, and
Heinrich R. Voth recorded the practices of the Hopi (Ortiz, 1979:
514-586).

Using this and other historical data and, informed by the
problems raised by the Thoms and their critics, I set out to establish the
nature of Hopi astronomical theory and to quantify Hopi astronomical practice. This would shed some light on the general nature of astronomy in
non-literate societies and establish the range of precision that we can
expect from naked eye horizon-based astronomies. Related to these archaeoastronomical problems are questions of whether the kind of markers used
to indicate sight lines among the Hopi would allow us to make inferences
as to the kind of markers used in the archaeological past.

But there are other aspects of the Hopi scientific community
that we must understand if we are to understand how it--like any other
scientific community--is able to perform its functions in society. How do

practicing sun watchers introduce apprentices into the concepts, practices, and traditions of that art in order to maintain its continuity?
How are detailed scientific knowledge and general cosmological principles
passed on from generation to generation? To answer such questions, we
must understand the essential core of such a society's scientific concepts, a core which might be quite different from our own.

THE SOLSTICES IN HOPI THOUGHT

Shortly after the summer solstice of 1893, Alexander M.
Stephen, who had then lived permanently among the Hopi for over two years
and had lived in the vicinity on and off since 1881, wrote to his mentor
J. Walter Fewkes of the Bureau of American Ethnography to recount his
discovery of such a conceptual difference:

> The Hopi orientation bears no relation to North
> and South, but to the points on the horizon which
> mark the places of sunrise and sunset at the sum-
> mer and winter solstices. He invariably begins
> his ceremonial circuit by pointing (1) to the
> place of sunset at summer solstice, next to (2)
> the place of sunset at winter solstice, then to
> (3) the place of sunrise at winter solstice, and
> (4) the place of sunrise at summer solstice, &c.
> &c.
> Doesn't that please you? . . . As soon as it
> flashed upon me, I hastened in to apply the key
> to some of the old fellows' knowledge boxes. And
> then they one and all declared how glad they were
> that I now understood, how sorry they had been
> that I could not understand this simple fact
> before (Stephen, 1893a).

This "simple fact" is truly a master key that unlocks many elements of
Hopi thought. The four cardinal directions of Hopi cosmology, and apparently those of many other American Indian cosmologies, are not the four
directions which the European tradition derives from an abstract geomet-
rization of space. Rather their cardinal directions are the empirically
observable ones defined by observations of sunrise and sunset at the
winter and summer solstices. The four solstitial directions not only
provide a stable empirical framework within which astronomical

Fig. 1   Directions Altar in Goat Kiva, 1891
          (Stephen, 1936)

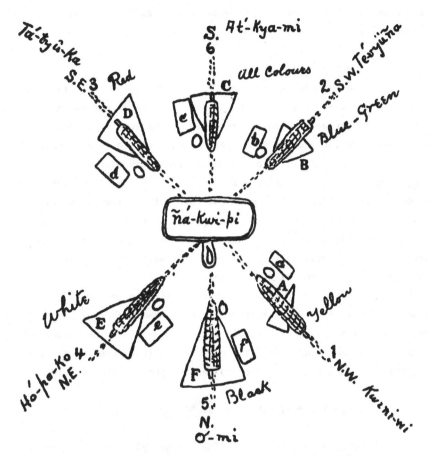

1. NW, Yellow corn ear, A. Oriole skin, a. unidentified bird
skin; 2. NW, Blue-green corn ear, B. Jay skin, b. Bluebird
skin; 3. SE, Red corn ear, D. Red-shafted woodpecker skin, d.
House finch skin; 4. NE, White corn ear, E. Whipporwhill skin,
e. Magpie feathers; 5. Above, Black corn ear, F. Blackbird
skin, f. Robin skin; 6. Below, Sweet corn ear, C. unidentified
bird skin, c. Warbler skin. (Bradfield, 1974, Stephen, 1936).

observations are made, but they also provide a general cosmological framework which draws apparently unrelated natural phenomena into an organic unity. This cosmic unity both governs and is portrayed in religious rituals, myths, and the like (Hieb, 1972: 84-89).

Sun Shrines

The solstice points, and especially the point in the southeast where the sun rises at the winter solstice and the diametrically opposite point in the northwest where the sun sets at the summer solstice are known as Sun's houses (Stephen, 1893c). At the edge of the mesas some ten kilometers across the valley to the southeast from each Hopi village there stood, and in some cases still stands, a small shrine called the Tawaki, or Sun's house. Such shrines mark the point of sunrise at the winter solstice and provide one of the fixed points of orientation for Hopi sun watchers (Fewkes, 1892, 1893). However, since they are small shrines, they cannot in themselves play the role of a distant foresight as they cannot be seen from the distant village.

It might be noted here that the shrines are small, easily disturbed, and bear few of the criteria to give them much credence as distant foresights, except that we are told that they are, and the direction in which they lie confirms this. The one element that does stand out is the sacredness of such sites. I took no photographs of them because

Table 1. HOPI DIRECTION SYMBOLISM

MSSS(NW)	MWSS(SW)	MWSR(SE)	MSSR(NE)	Above	Below
Yellow	Blue/ Green	Red	White	Black	All colors
Oriole	Mountain Bluebird	Parrot/ Macaw	Magpie	Cliff Swallow	Canyon Wren
Corn	Beans	Squash	Cotton	Watermelon	All Kinds
Mountain Lion	Black Bear	Gray Wolf	Wildcat	Golden Eagle	Badger
Deer	Mountain Sheep	Antelope	Elk	Jack Rabbit	Cottontail

After Hieb (1972), Levi-Strauss (1966), and Stephen (1936).

photography would have been sacrilegious. The Hopi friend who led me to
some of them immediately paused to pray upon our arrival. If the Hopi
case is any indication, one factor common to archaeoastronomical sites may
be their reputation as holy places.

Yet for the Hopi, this sacredness plays a role in passing on
elements of their astronomical tradition. Each year when the Sun arrives
at his house, either the winter house to the southeast or the summer house
to the northwest, prayer sticks (or Pahos) adorned with feathers and other
ritual symbols are made to be offered to the Sun (Crow Wing, 1925: 95;
Parsons, 1939: 393-4; Stephen, 1936: 60-61; Titiev, 1944: 146, 299).
During the period of four days when the Sun is said to stay at his house,
these offerings are deposited at the shrine by one of the younger members
of the society responsible for the solstice ceremonies. Charging a young
man with responsibility for a ten kilometer run across the desert seems
reasonable, yet this element of ritual also serves to instruct him of the

Fig. 2.   Sun Prayer-Offerings
          (Stephen, 1936)

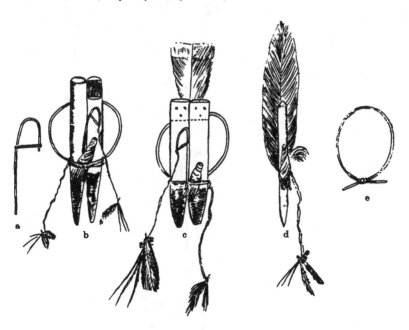

a. Age Crook, b. & c. Sun Pahos, d. Warrior Paho, e. Sun
House.

Fig. 3
(Stephen, 1936)

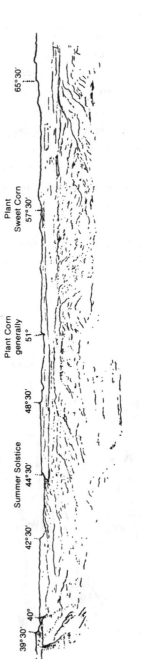

visible landmarks near the Sun's house which, although it marks one of the
cardinal points of Hopi astronomy and cosmology, cannot be seen from the
village.

A more convenient although less precise point of orientation
is provided by the sacred mountains in the direction of winter solstice
sunset, the San Francisco Peaks. Almost anywhere one goes in Hopi
country, these snow capped mountains can be seen to the southwest, generally surmounted by a nimbus of clouds. And to the Hopi, these mountains
are further sanctified by being the home of the Kachinas, who bring rain
to the Hopi people and their crops.

Solar Observations and Crops

The sacred system of solstitial directions plays a major role
in maintaining the Hopi agricultural cycle. As the sun moves northward
along the eastern horizon, his rising is watched by the tawa-mongwi, or
Sun Chief, who notes the sun's arrival at a series of distinctive natural
landmarks (Stephen, 1893b, 1936). He then announces when it is time to
begin the plantings of the various crops, beginning with sweet corn which
is planted in fields in sheltered nooks where the radiated heat from the
mesa walls protects it from nighttime frost, and ending with the planting
of the principal corn crop, which is planted from May 21st to mid June,
although later plantings to replace seeds that fail to sprout continue
until the sun leaves his summer house.

This planting sequence provides a clear indication of how
carefully the Hopi can relate their observation of the sun to their practical needs and to their observations of other natural phenomena. If we
look at the weather records from the nearby weather station at Keams
Canyon (which was active from 1906 to 1928 and from 1953 to the present)
we find that there is about a fifty per cent chance of freezing temperature on or after May 21st, and ten days later, when the corn plants begin
to sprout, the chance of a subsequent freeze is about twenty per cent
(U.S. Weather Bureau, 1931, 1949-77).

Even the end of late plantings around the summer solstice is
not arbitrary. There are about 107 days from the solstice to the date on
or before which there is a fifty per cent chance of freezing temperature
and 120 days to the probable date of a hard freeze of 28⁰ F. Since Hopi
corn matures in 115 to 130 days, there is usually enough time in Autumn
for even the latest plantings to reach full maturity.

The Hopi themselves recognized the relation between careful observation of the sun, the maintenance of the calendar, and the effect of frost on the coming crops. Crow Wing, speaking of a later ritual observation for the Snake Dance said:

> Sometimes, if the Snake Chief does not watch the
> sun right, they dance early, then it freezes early
> too. That is the reason why they must try and
> watch the sun very closely (Crow Wing, 1925:
> 101).

Similar concerns apply to the observation of sunrise for planting in the Spring.

> We think the Sun-watcher is not a very good man.
> He missed some places, he was wrong last year
> . . . . All the people think that is why we had
> so much cold this winter and no snow (Crow Wing,
> 1925: 74-75).

It is not only the experts who watch the sun, but as these passages show, even the common Hopi farmer can criticize a qualified sun watcher. Clearly, the Hopi have empirically worked out and disseminated a precise relationship between the motion of the sun, their local climate, and the

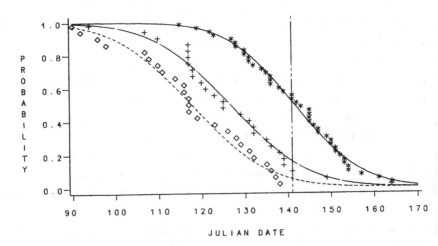

Fig. 4 PROBABILITY OF SPRING FREEZE
AFTER GIVEN DATE — KEAMS CANYON, ARIZ.

* = 32°F, + = 28°F, diamond = 24°F.

primary concern of an agricultural people, the growth of their crops. As
we will see later, this concern provides one of the central themes of Hopi
religion.

Solar Observations and Ritual

At the summer solstice the Sun is said to stay at his summer
house in the northwest—directly opposite his winter house in the southeast. When the Sun leaves his summer house the pattern of sun watching
changes in several ways. No longer is the observation of the sun concerned with regulating the agricultural cycle, rather it is concerned with
regulating the ceremonial aspects of the year. Secondly, observations are
now made at sunset as the sun moves south along the western horizon rather
than at sunrise. Finally, the observations are not made by the Sun Chief
proper but by an elder of that clan principally responsible for the forth-
coming ceremony. The Powamu Society Chief, from the Kachina Clan, watches
the sun for the homegoing of the Kachinas or Niman in July; the Singers'
Society Chief, from the Tobacco Clan, watches sunset for the Tribal Initiation or Wuwuchim in November, and finally the Sun Chief again watches the
sun for the Winter Solstice ceremony or Soyal.

According to somewhat conflicting reports, he will begin
watching sunset either a given number of days after the preceding ceremony
or when he notices that the sun is setting near the proper place. The
point that marks when to begin watching need not be traditional. One sun
watcher said that he began watching when the sun "went in" near the cultural center and motel on Second Mesa (McCluskey, 1978-80). After this
distinctly un-traditional beginning, he would then get out his ritual
paraphernalia, sit at a specific place in the village (which would be
different than the places that other elders used in watching for their
ceremonies) and watch until the day that the sun set at the proper point
on the horizon. Knowledge of these precise observation rituals had been
passed down to him within his clan from his uncles, and from their uncles
before them. The evening that the sun reached the proper point, he would
then meet with the other elders belonging to his ceremony, and they would
smoke together and begin preparing themselves spiritually for the coming
ritual.

It is quite apparent that these observers are not the compara-
tively disinterested observers of the heavens that we find reflected in
modern catalogs of star positions. Yet their observations are not any

less precise because they are carried out as part of religious rituals.
In fact the conservatism of ritual seems to account for the care with
which these observations are carried out.

Observational Precision

Given the date on which a festival occurred we can determine
the longitude of the sun on the day of that festival and, from the known
interval from observation of the sun to the festival, the longitude of the
sun on the day of the observation. Taking a series of festival dates, we
can determine the standard deviation of a single observation of the longitude of the sun. Assuming a uniform horizon, this standard deviation in
longitude can be converted to a corresponding deviation in azimuth. Since
there is considerable ambiguity in the descriptions of the methods used to
watch the sun, this internal comparison of these festival dates offers a
better indication of observational precision than comparison of festival
dates with some theoretical ideal derived from geodetic calculations.

Before going further, some comments about the sources of error
influencing observations of sunrise and sunset seem in order. One factor
derives from the very nature of horizon observations. Horizon observations are of the sort "yes, the sun has set at the proper point today" or
"no, it has not yet reached the proper point to begin the appropriate
ceremonies". These are observations that can only be made once every
twenty-four hours--just as the resulting religious ceremonial can only
be held on one day or the next. This introduces a quantum effect into the
measurements corresponding to the sun's movement in twenty-four hours.

The remaining sources of error due to atmospheric refraction,
the personal equation of the observer, or even changes in alignment of the
instrument (say--if trees grew in a valley forming a distant foresight or
the house from which observations were made was remodeled) comprise what
we would usually consider observational error.

The variation due to the sun's daily motion can be computed on
the basis of a uniform rectangular distribution and this variation can be
subtracted from the variation in the observations. This leaves the variation due to observational error only and the corresponding standard deviation of an individual observation can be computed in the normal manner.
As it turned out, in this case the quantum effect due to the sun's daily
motion only made changes of a few percentage points in the standard deviation. In cases where the sun moves rapidly along the horizon, such as

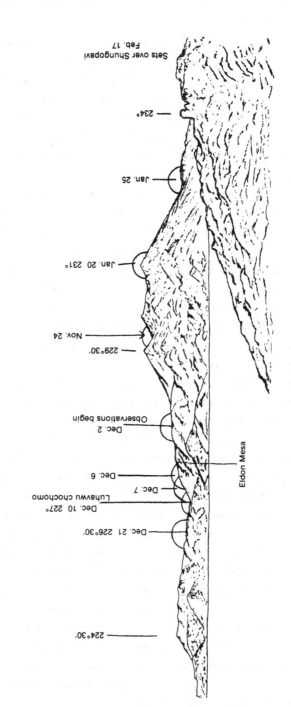

Fig. 5
(Stephen, 1936)

Walpi Western Horizon
(magnetic bearings from Red Cape / observations from Bear clan house)

equinox observations in near polar latitudes, this day to day change would
play a greater role.

Since the Hopi festivals were based on observations near the
solstices, this factor was not significant, but it was corrected for none-
theless. The standard deviation in azimuth that emerged was, with one
exception, in the range ±20' of arc. The festival dates at Walpi
Pueblo for the Tribal Initiation (Wuwuchim) in November and the Kachina
homegoing (Niman) in midsummer and the dates at Oraibi pueblo for the
Kachina's return (Soyal) at winter solstice all reflected horizon observations with this range of precision.

The one exception was the Kachina's return (Soyal), at Walpi.
Here the observation, made 11 1/2 days before the winter solstice, was
accurate to ±10' of arc, but this is an exceptional observation. The
baseline is 127 kilometers long through clear desert air (I would note
that the Lowell Observatory and a branch of the U.S. Naval Observatory
were built near one end of this sight line because of the exceptional
seeing). Furthermore, the sight line begins some 150 meters above the
desert floor and, due to the intervening valley of the Little Colorado
river, continues at even higher elevations through most of its length.
This reduces the refraction effects to which sight lines near the ground
are prone. It is also possible that a count of days from the earlier
observation for Wuwuchim, when the sun was moving more rapidly, guided the
sun watcher in his observations of the more slowly moving sun for Soyal.

The observational error for Soyal at Walpi is less than that
which would be introduced by the sun just touching the horizon, bisected,
or just below the horizon, and there is no clear ethnographic evidence as
to how the Hopi defined sunset. The record of Soyal dates, however,
suggests that the Hopi defined sunset as either when the sun's image was
bisected or perhaps when the setting sun first touched the horizon.

Given the Hopi's ritual concern that sun watchers observe
carefully in order to get the observation right, it seems that this fairly
well represents the maximum precision we can expect from careful naked eye
horizon observations of the sun. Being measured along the horizon, the
reference against which observations are actually made, this benchmark in
azimuth can readily be transferred to sites at other latitudes. Such a
benchmark suggests that we should carefully examine any archaeoastronomical hypotheses that require individual horizon observations with precision exceeding ten minutes of arc.

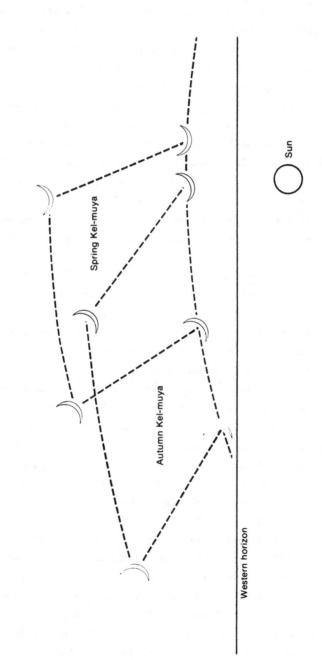

Fig. 6

Varied appearance of New Moon at
Spring and Autumn Kel-muya

THE MOON AND THE CALENDAR

In addition to these solar observations, at least two forms of observations of the new moon have been documented among the Hopi. On the one hand, there are qualitative observations of the New Moon that are associated with characteristic weather patterns. While observing the new moon of Kel-muya in the spring of 1893, Stephen was told that "When the crescent is on its back it is called horizontal and is not good; when vertical, it is good and it will be moist" (Stephen, 1936: 389). This reflects the seasonal differences between the two annual occurrences of Kel-muya. In spring, when the ecliptic is almost vertical on the Western horizon at sunset, the new moon first appears lying on its back, and spring in Hopi country is dry and windy. In autumn, the new moon is tilted on its side, and seasonal rains occur in late summer and early fall, insofar as Northern Arizona can be said to have a rainy season.

The Year's Duality

As with other Indians of the Southwest, the other form of lunar observation serves to find the beginning of a month in the solar calendar. The names of the months in the Hopi calendar display one striking feature reflecting the year's duality in Hopi cosmology. Although there is considerable confusion as to the names of the months from May to October, the earliest evidence shows that it was common to refer to the five summer months by the same names as the five winter months. The wide distribution of this pattern among the Western pueblos and the Indians of Southern California is illustrated by Figure 7. The Hopi would illustrate this identity of the summer and winter months by laying the five fingers of the right hand atop the five fingers of the left hand, naming the five months of the upper hand and saying that the five under fingers were the same (Stephen, 1894a, 1894b).

This duplication of the names of the months appears to reflect a fundamental duality in the Hopi conception of time. Sakwistiwa, an elder at Walpi pueblo said that when the winter Powa-moon is shining in the Above, (that is, in the world where we live), its counterpart the Nashan-moon or summer Powa-moon, is shining in the Below. When the Nashan shining in the Below comes to the Above in September, it will bring just the same harvest that it is now shining on in the Below (Stephen, 1894c; Titiev, 1944: 173-176).

Fig. 7   Types of calendars in the southwest
         (after Spier, 1955; courtesy Museum of Northern
         Arizona)

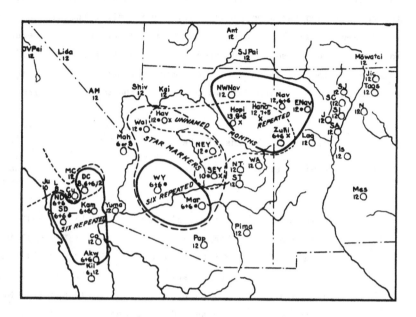

Legend:

Calendar elements: (12, 8+5, etc., number of months; •,
stellar indicators; x, unnamed months).

Tribal names and abbreviations: Akw, Akw'ala; AM, Ash Meadows;
Ant, Antarianunts; Co, Cocopa; Cu, Cupeno; DC, Desert
Cahuilla; Hano, Hopi-Tewa; Hav, Havasupai; Hopi; Is, Isleta;
J, Jemez; Jic, Jicarilla; Ju, Juaneno; Kai, Kaibab; Kam,
Kamia; Kil, Kiliwa; L, Luiseno; Lag, Laguna; Lida; Mar,
Maricopa; MC, Mountain Cahuilla; Mes, Mescalero; Moh, Mohave;
Mowatci; N, Nambe; Nav, Navajo; ND, Northern Diegueno; NEY,
Northeastern Yavapai; NT, Northern Tonto; NWNav, Northwestern
Navajo; OVPai, Owens Valley Paiute; Pap, Papago; Pima; SA,
Santa Ana; SC, Santa Clara; SD, Southern Diegueno; SEY,
Southeastern Yavapai; Shiv, Shivwits; SI, San Ildefonso; SJ,
San Juan; SJPai, San Juan Paiute; ST, Southern Tonto; Taos;
WA, Western Apache; Wal, Walapai; WY, Western Yavapai; Yuma;
Zuni.

Lunar Observations and Ritual

Watching the new moon seems to have been a regular occurrence, but it is only in anticipation of this moon, the Powa-moon, that watching for the new moon takes on ritual importance. After the winter solstice the Powamu Chief--who is also the sun watcher for the Niman Ceremony which follows the summer solstice--watches the moon. He begins his private ceremonials looking forward to the coming new moon on the evening of the Pa-muya full moon, which follows the end of the winter solstice ceremony.

Table 2. Months of the Hopi Calendar

Hopi Name(s)	Translation	Range of Dates for New Moon
Kele-muya	Initiates' moon	15 Oct - 12 Nov
Kya-muya	Dangerous moon	13 Nov - 11 Dec
Pa-muya	Moisture moon	13 Dec - 10 Jan
Powa-muya	Purification moon	11 Jan - 8 Feb
Isu-muya	Cactus blossom moon	10 Feb - 10 Mar
Kwya-muya	Greasewood fence moon	11 Mar - 8 Apr
Hakiton-muya	Waiting moon	10 Apr - 8 May
Uusu-muya*	Planting moon	9 May - 6 June
Kele-muya	Initiates' moon	20 May - 16 June
Niman-muya	Homegoing moon	7 June - 5 July
Kya-muya	Dangerous moon	18 June - 16 July
Pa-muya	Moisture moon	18 July - 15 Aug
Nashan-muya	Big feast moon	17 Aug - 14 Sept
Powa-muya	Purification moon	
Tuhoosh-muya	Basket moon	
Angok-muya	Harvest moon	15 Sept - 13 Oct
Isu-muya	Cactus blossom moon	

* One intercalation hypothesis that reconciles most of the available data is that the non-repeated series of months, beginning with Kwya-muya, does not mesh perfectly with the succeeding months of the repeated sequence. Thus Uusu-muya would sometimes fall before Kele-muya and represent an intercalary month, but would usually coincide with Kele-muya. The place of the summer Pa-muya in the calendar is fixed by the record of the dates of the Snake and Flute ceremonies. The dates of the summer Kele and Kya-muya are probably synchronized with the subsequent Pa-muya.

In the 1880's Intiwa, the Powamu chief at Walpi, explained this practice
by saying:

> When my people had learned to build houses and men
> had grown accustomed to life in the [kivas], Masau
> came and taught them many things concerning growth
> of plants and trees and instructed them about
> planting beans when the moon should be at a cer-
> tain age and after the sun had come a certain dis-
> tance on his way back to the north. Many, many
> days this has been the custom and we have no right
> to forsake the ways of our fathers (Stephen, 1883:
> 60-61).

RITUAL TRANSMISSION OF ASTRONOMY
Esoteric Ritual and Astronomy

Two weeks later, when the Powamu chief saw the first crescent
of the Powa-moon, he would call the elders of the Powamu society together
to hold a "smoke talk", a prayerful ritual to pray for, and confirm the
date of, their coming ceremony.

This ritual was most completely recorded in the 1890s at
Oraibi pueblo by Rev. H. R. Voth. As the smoke talk progressed, a number
of songs were sung, one of which consists of twenty variations of the
following verse:

> The sun he is bringing
> The sun he is watching
> When at Apoonivi the sun is setting
> The plants are being clothed (Voth, 1901).

The first ten verses name the points where the sun sets on the Western
horizon during the planting season, the next ten verses name corresponding
points of sunrise on the Eastern horizon.

Here we have a striking example of a ritual that is connected
with the coming growth of crops that also provides the participants with a
recitation of the points to watch the sun for the coming planting season.
By naming the points in ritual, the accurate transmission of this knowledge to future generations is made more likely.

Popular Ritual and Astronomy

After the smoke talk, the chief of the Powamu society
announces that it is time for everyone to begin planting beans as a ritual

in the underground kivas. The kivas will be heated and the beans forced
to germination to be distributed later in the ceremony in anticipation of
the crops that will grow later in the year. The ritual planting is
intended to forecast the fruitfulness of the coming harvest in a way that
reflects the duality of the Hopi conception of time. The beans sprouting
in the underground kivas at Powamu in some way correspond to the harvest
that already exists in the Below and that will come to fruition this
autumn in the Above. To announce the time for this planting, the Powamu
chief distributes ritual offerings to each kiva in the village. These
offerings are prayer sticks to which is fastened a blackbird feather, and
Rev. Voth was told that the Powamu chief is said to represent the blackbird, Tokutska, whose arrival signals the time to plant crops.

        This combination of ritual and nature lore doesn't seem imme-
diately connected with astronomy, but the point where the sun rises to
mark the early planting for the pueblo of Oraibi is called Tokutska
Masaat, Blackbird's Wing, because as one farmer said "whenever [the
blackbirds] come, they know that it's planting time . . . . The sun is
going--and where the sun comes up--and the same time, that's when the
birds come, that's where they named it Tokutska" (McCluskey, 1978-80).
Thus the blackbird, who arrives in May, provides a connection between the
profane agricultural aspects of Hopi astronomy and its sacred ritual
aspects.

        Initiation Ritual and Astronomy
        The Powamu Ceremony is not just a preparation for the coming
growing season, as Voth noted it also includes an initiation ceremony
during which some of the secrets of Hopi ritual are revealed to young
children. On the fourth day of the ceremony the initiates witness a dance
by the Chowilawu Kachina, in which he dances four times counterclockwise
around a sand mosaic, each corner of which is marked by the color of the
appropriate direction. On the remainder of the mosaic are five multi-
colored blossoms and a scattering of many variously colored dots, all said
to represent the blossoms of newly growing plants. Recall that Hopi symbolism relates all colors to the Below, the direction from which crops
come and where they are already growing. These multi-colored blossoms
exist now in the Below and will come to the Above in the future.

        But even this is not free of astronomical content. As
Chowilawu dances, he displays a circular book-like object, containing six

Fig. 8   Disks carried by Chowilawu Kachina
(after Voth, 1901)

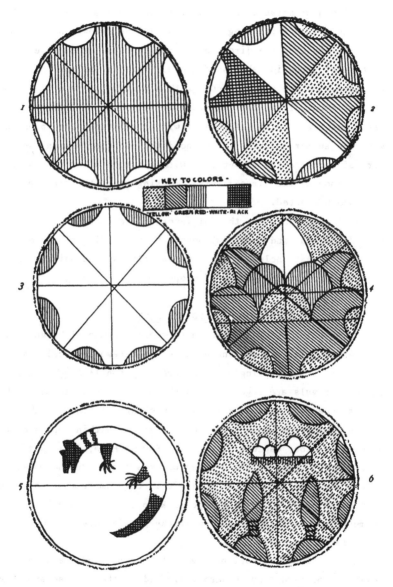

1. Red blossom, 2. Multi-colored blossom, 3. White blossom, 4.
Sprouting leaves, 5. Moon symbol, 6. Corn plants and rain
clouds.

circular pages, each bearing a painting whose symbolism relates astronomy
to the growth of plants. First is a red blossom, representing the direction of winter solstice sunrise, followed by a multi-colored blossom, like
those in the sand painting, flowering in the Below. Next is a white blossom, representing the direction of summer solstice sunrise, followed by a
picture of sprouting shoots of green. This is followed by a symbol for
the moon, and the sequence ends with a picture of rain clouds and growing
ears of corn. If we consider them in pairs, we see germination in the
Below associated with the time of winter solstice, the sprouting of plants
with the summer solstice which ended the planting season, and the growing
ears of corn and the summer rain with the moon after the solstice, which
is the moon of the Niman festival, during which the first ears of sweet
corn are harvested.

In this way the initiates are introduced to the four solstitial directions and the relationship of these directions to a complex set
of symbols that include colors, birds, plants, and animals and an intimate
involvement with the growth of crops.

RELATING THE SUN AND THE MOON

The two festivals in the Hopi ceremonial calendar that I have
ignored thus far are the Snake and Flute ceremonies, celebrated on alternate years. Interestingly, there are many conflicting reports as to how
these festivals are regulated and the only thing I can say with certainty
is that the date varies, perhaps with changes in the length of the growing
season, sometimes because of social tensions within Hopi society, although
at some pueblos, especially those on Second Mesa, it is sufficiently well
behaved to suspect possible solar observations. Furthermore, a definite
attempt is made to keep the date within the summer month of Pa-muya
(McCluskey, 1981). This attempt to keep the Snake and Flute ceremonies in
Pa-muya is a further indication of the Hopi's interest in relating lunar
and solar phenomena.

That the Hopi observe both the sun and the moon and note the
relations between them is also indicated by the luni-solar principles
underlying Powamu, and by the recognition that while in most cases specific festivals occurred in specific months, there are exceptions to the
rule.

Fig. 9

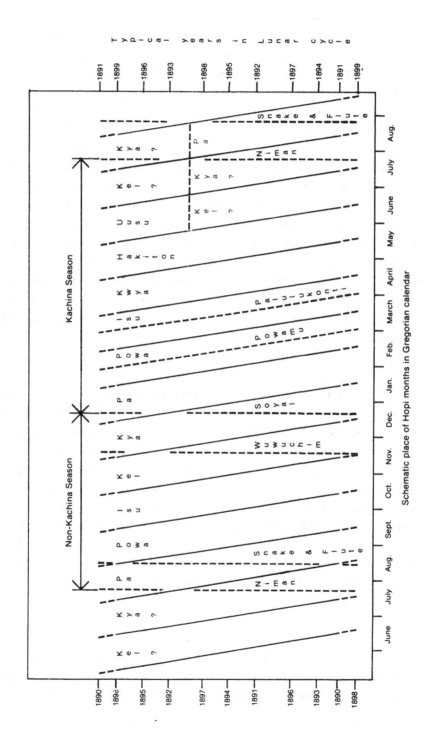

Sometimes the Sun Goes Slowly

In the 1920s Elsie Clews Parsons noted that the winter Pa-muya is so named after the games played in that month and that Soyal generally occurred in the preceding month Kya-muya. But sometimes "if the sun goes slowly and Soyala comes late, Soyala may come in pah muye, but even so they would not open the games of pah muye until four days after Soyala" (Stephen, 1936: 239).

The instances when Soyal occurs in Pa-muya are explained on the basis that at such times the sun is going slowly. It is apparent in this case that the reference frame is the lunar month, which is taken as proceeding uniformly. A Hopi neighbor of mine one summer asked me, since I claimed to know something about astronomy, whether the moon or the sun—— that is the calendar month—was going more slowly. After my confused reply, he told me the moon was going about ten days slower, which agreed with the fact that in 1980 the astronomical new moon had occurred on July twelfth and August tenth. Here we have another instance of a Hopi noting the current difference between a lunar and solar calendar, but here the amount by which the moon goes fast or slow is counted.

Epact

This concept must be an indigenous one, for the sun or moon never go fast or slow in European astronomy. But a related concept is found in European calendrical astronomy, that of the epact which is defined as the age of the moon on a specific date, or more properly as the age of the moon on New Year's Day. Although they are based on different cosmological principles, the epact and the Hopi estimate of how much the sun is going fast or slow contain the same calendric information. If you know the epact for a given year, you can compute the date of all the new moons and the corresponding dates of all the religious ceremonies for that year. Similarly, if Wuwuchim comes late, so will Soyal, Niman, the Snake Dance and the agricultural cycle of plantings come late in reference to the lunar calendar.

Computus at Zuni?

Since some Hopi kept track of how much the sun was going fast or slow, is there any indication that they could make year-to-year predictions of this phenomenon, using methods analogous to those employed by

medieval practitioners of the ecclesiastical art of computus. There is
one tantalizing account from the neighboring pueblo of Zuni that suggests
this may be the case.

At Zuni Ruth Benedict (1935, II: 62-68) collected a version of
the Snake Myth, a myth common at Hopi as an explanation of the origin of
the Snake Clan, which was used at Zuni to explain why the Spider Clan held
the office of Pekwin or sun-watcher. This Zuni version recounts how the
young hero, after visiting the Snake people in the West then went on to
visit the Sun (and in some versions, the Moon also) at his Western house.

> The Sun told him 'When you get home you will be
> Pekwin and I will be your father. Make meal
> offerings to me. Come to the end of the town
> every morning and pray to me . . . . At the end
> of the year when I come to the South, watch me
> closely; and in the middle of the year, in the
> same month, when I reach the farthest point on the
> right hand, watch me closely.' 'All right' [the
> youth replied]. He came home and learned. The
> first year at the last month of the year he
> watched the Sun closely, but his calculations were
> early by thirteen days. Next year he was early by
> twenty days. He studied again. The next year his
> calculations were two days late. In eight years
> he was able to time the turning of the sun
> exactly.

Now there is one element in this passage that is quite unusual
among Pueblo myths. In most Pueblo tales, numbers, especially for duration, are given ritually significant values: four, eight, or sometimes
sixteen or twelve days are common. In this passage the only ritually
significant number is the eight years that pass until the Pekwin's calculations are exact. But it turns out that these other numbers approximate
values significant to the Pekwin's observations. If we consider an eight
year cycle through which the epact runs (corresponding to the octaeteris
of the Greek tradition) and compare that with the values in this tale, we
find a rough correspondence between how early or late the Pekwin's calculations of the solstice (presumably based on a lunar calendar) were and
how early or late the moon would be in a simple eight year cycle.

This myth admittedly provides ambiguous evidence from which to draw an astronomical theory. In fact Jonathan Reyman (1980), following Ruth Bunzel (1932: 511-13), took it as evidence of the limited ability of the Zuni Sun priests to reconcile the lunar and solar calendar. Yet this passage could very well be a fragment of esoteric lore dealing with the reconciliation of the lunar and solar calendar that Benedict's informant-- a knowledgeable, but unstable figure--inadvertently placed in his recita- tion of the widely known Hopi Snake Myth (Benedict, 1935, I: xxxviii-xl).

In that case, it would represent a collective attempt to deal with the perceived irregularities of the Sun's motion. If, on the other hand, it merely represents the individual contribution of Benedict's Zuni informant, it certainly represents his personal attempt to deal with these irregularities.

Furthermore, such an account fits well with what we know from the Hopi of their interest in the sun going fast or slow, and would be admirably suited to empirical correction. If the sun is seen to be thir- teen days fast this year, the myth suggests that it would be twenty days fast next year. This corresponds to an interesting fact I was told by one Hopi. The Sun Clan is not so much interested in the habits of the sun as in why it goes slow sometimes. In looking at the astronomies of other cultures we should not let our conviction of regularity and periodic cycles blind us to what, in other scientific frameworks, are perceived as very real irregularities--irregularities which their investigations must attempt to account for.

Table 3. THE SNAKE MYTH AND THE LUNAR CYCLE

Pekwin's "calculations" of solstices	Eight-year cycle of epact
. . .	-.8 d
13 days early	-11.7 d
20 days early	-22.5 d
2 days late	-3.9 d
. . .	-14.8 d
. . .	-25.6 d
. . .	-7.0 d
. . .	-17.9 d
exact	+.8 d

ASTRONOMY WITHOUT ASTRONOMERS

I have only given a brief survey of Hopi astronomy, but I hope that some of the flavor of it may have come through, and it seems to me that it suggests a paradox. Considered as astronomy, it shows all the concern with exact observation and the development of observational and theoretical frameworks that we would expect of modern astronomy. The evidence regarding the sun going fast or slow even suggests the Hopi were able to predict whether a future ceremony would be early or late, a falsifiable prediction that would seem to satisfy even Sir Karl Popper's rigorous criterion of science.

Yet, while it seems that there is a Hopi astronomy, there are certainly not Hopi astronomers. Hopi astronomy keeps making contact with Hopi meteorology, Hopi agriculture, Hopi theology, and other elements of Hopi thought precisely because the individuals involved in Hopi study of the skies are also actively concerned with weather lore, farming, and religion.

The Hopi have no real professional astronomers, just as they have no narrow specialists in meteorology, agriculture, or theology. Instead they have elders, widely educated in the ritually transmitted wisdom of clan and tribe.

ACKNOWLEDGEMENTS

I would like to thank the many archivists, librarians, and members of the Hopi tribe who helped me in many ways, and the National Science Foundation for a grant (SOC 78-05768) that made my archival and field research possible.

REFERENCES

Benedict, R. (1935). Zuni mythology, 2 Vols. (Columbia University
  Contributions to Anthropology, 21). New York.
Bradfield, M. (1974). Birds of the Hopi Region. (Museum of Northern
  Ariz., Bull. 48). Flagstaff, Ariz.
Bunzel, R.L. (1932). Introduction to Zuni ceremonialism. Bureau of
  American Ethnology, Annual Report 47, 467-544.
Crow Wing. (1925). A Pueblo Indian journal, (ed.) E.C. Parsons, (Mem.
  Amer. Anthr. Assoc., 32). Menasha, Wis.
Fewkes, J.W. (1892). Awatobi field notebook. Washington, Smithsonian
  Institution, National Anthropological Archives (SI/NAA), File
  4408(13).
────── (1893). A-wa-to-bi: Archaeological verification of a Tusayan
  Legend. Amer. Anthr. 6, 363-375.
Hawkins, G.S. & J.B. White. (1965). Stonehenge Decoded. Garden City, NY:
  Doubleday & Co.
Hieb, L.A. (1972). The Hopi ritual clown: Life as it should not be.
  Ph.D. Dissertation, Princeton University, Princeton, N.J.
Levi-Strauss, C. (1966). The Savage mind. Chicago: University of
  Chicago Press.
Lockyer, J.N. (1906). Stonehenge and other British stone monuments
  astronomically considered. London: MacMillan & Co.
McCluskey, S.C. (1978-80). Hopi Field Notes.
────── (1981). Transformations of the Hopi calendar. In Archaeoastronomy
  in the Americas, ed. R.A. Williamson, pp. 173-182. Los Altos,
  California: Ballena Press.
Ortiz, A. (ed.). (1979). Handbook of North American Indians, Vol. 9,
  Southwest. Washington: Smithsonian Institution.
Parsons, E.C. (1939). Pueblo Indian Religion. Chicago: University of
  Chicago Press.
Reyman, J.E. (1980). The Predictive dimension of priestly power. In New
  Frontiers in the archaeology and ethnohistory of the greater
  Southwest, ed. C.L. Riley & B.C. Hedrick. (Trans. Ill. State
  Acad. Sci. 72, 4). Springfield, Ill.
Spier, L. (1955). Mohave culture items. (Museum of Northern Ariz.,
  Bull. 28). Flagstaff, Ariz.
Stephen, A.M. (1883). Hopi notebooks, Vol. 1, Legends. Columbia Univer-
  sity Library. Elsie Clews Parsons Collection.
────── (1893a). Letter to [J.Walter] Fewkes, June 29, 1893. Washington,
  SI/NAA, File 4408(4).
────── (1893b). Letter to [J.Walter] Fewkes, Nov. 30, 1893. Washington,
  SI/NAA, File 4408(4).
────── (1893c). Letter to [J.Walter] Fewkes, Dec. 21, 1893. Washington,
  SI/NAA, File 4408(4).
────── (1894a). Letter to [J.Walter] Fewkes, Jan. 11, 1894. Washington,
  SI/NAA, File 4408(4).
────── (1894b). Letter to [J.Walter] Fewkes. Jan. 18, 1894. Washington,
  SI/NAA, File 4408(4).
────── (1894c). Letter to [J.Walter] Fewkes, Saint Valentine's Day.
  Washington, SI/NAA, File 4408(4).
────── (1936). Hopi Journal of Alexander M. Stephen, ed. E.C. Parsons.
  New York: Columbia University Press.
Thom, A. (1967). Megalithic sites in Britain. Oxford: Clarendon Press.
────── (1971). Megalithic Lunar observatories. Oxford: Clarendon Press.

Thom A. & A.S. Thom. (1978). Megalithic remains in Britain and Brittany.
        Oxford: Clarendon Press.
Titiev, M. (1944). Old Oraibi: A Study of the Hopi Indians of Third
        Mesa. (Papers of the Peabody Museum of American Archaeology
        and Ethnology, XXII, 1). Cambridge, Mass.
United States Weather Bureau. (1931). Climatic summary of the
        United States, Bulletin W, Section 25, Northern Arizona.
        Washington.
United States Weather Bureau. (1949-77). Climatological data, Arizona.
        Washington.
Voth, H.R. (1901). The Oraibi Powamu ceremony. (Field Columbian Museum,
        Anthr. Ser., Vol. 3, no. 2). Chicago.

THE SIDEREAL LUNAR CALENDAR OF THE INCAS

R. T. Zuidema
University of Illinois, Urbana, Illinois 61801, U.S.A.

The Incas were the last exponents of a long tradition in Andean cultures that the Spanish conquerors in the sixteenth century became familiar with. They are the only native American people about whom we possess detailed information on concepts of astronomy and calendar that allow us to study its characteristics of precision needed in a state bureaucracy highly concerned with recording and correlating cycles of irrigation, agriculture, husbandry, trade and warfare. These operated within an intricate system of kinship, age classes and socio-political organization. The Spaniards were acute observers and they left us extensive data on Inca culture. However, when they detached data on Inca history, political organization or calendar from myth and ritual, that is from the way the Incas expressed themselves on their own culture, their Western interpretation and interests became too prominent and obscured the data. Only with the study of myth and ritual can we reconstruct the original and precise aspects of the Inca calendar.

Cuzco, the capital of the Inca empire, is at the upper West end of an intermontane valley, 13½° South of the Equator, about 20 kms long and some 3300 m. high. Just above the city begins the river Huatanay that flows in a direction of some 15° South of East. After it has passed another valley it flows in the Vilcanota or Urubama river that goes in a northwesterly direction. The data to be used here on the Inca calendar, myth and ritual all refer to the valley of Cuzco and are correlated to a system of geographical description for which the Incas used 41 directions as considered from the central temple of the Sun in Cuzco, named Coricancha. As far as we have been able to study these directions, called ceques, "lines", they were based on sightlines toward the horizon. Ceques to the nearby horizon could pass beyond it, while ceques to the far away horizon would end before. The directions were known with the help of natural or man made markers along the ceques, in numbers varying

from 3 to 15, whose locations normally were chosen as close as possible
to the directions. For this reason they were worshipped as sacred and
called huacas. There were many reasons for places to be sacred; the
huacas of the ceque system were neither the only nor necessarily the most
sacred ones. Important huacas could happen to be used in the ceques, but
also insignificant huacas can provide much needed precise information.
Ceques formed a system of coordinates by which information of very different orders was organized, like is done in our maps (Zuidema 1964, 1977,
for detailed description of the system).

A SUMMARY OF THE ARGUMENT CONCERNING THE RECONSTRUCTION OF
THE CALENDAR

The astronomical phenomena used for constructing the Inca calendar were the sun, the moon and certain stars, in particular the
Pleiades as the "mother of the stars". Although the Incas may have named
far more stars and constellations, at the moment we only have available
to us, from the Incas and from modern Andean villages near Cuzco (Urton
1981) systematic data on stars and constellations along the Milky Way and
on the dark patches of interstellar dust in the Milky Way recognized as
"dark cloud" constellations. Of special importance among the latter were
a black Llama in the Milky Way with α and β Centaurus as her eyes, and
the "Coalsack", in the Southern Cross, known as Yutu or, in present day
Peruvian Spanish, "partridge". The worship of Venus was important in
relation to the Inca atmospheric system, but we do not have data yet on
any calendrical function.

The Incas were interested in the correlations of sun to moon,
of sun to stars and of moon to stars. They measured key events of these
three cycles—with the help of quipus, bundles of knotted chords representing numbers—in terms of a sidereal lunar year of 328 nights (12 x
27 1/3 = 328) with a fixed position in the solar year. The Inca calendar
as a political instrument was primarily sidereal lunar.

With two exceptions around the December solstice, the Incas
named synodic months after seasonal activities. The month of first
planting was tied to observing sunset on August 18. Knowledge of how
many days full moon would occur in a specific year before or after August
18, with an observed limit of 15 days on either side, enabled the Incas
to calculate how early or late other months were in a specific year and
how to adjust their seasonal activities to the lunar sequence. The Incas

began their count of synodic months with the one during which the June
solstice occurred. The earliest visible moon of such a month falls on
May 26. Sunrise on that day had particular significance for calculating
the moon of **Inti raymi**, "the feast of the Sun".

The count of the sidereal lunar year of 328 nights started on
the night of June 8-9 and ended on that of May 3-4. The starting date
was chosen in relation to the first heliacal rise of the Pleiades around
June 5-8, to the first full moon after and to the June solstice. We only
have explicitly stated evidence to establish a sun-moon-star correlation
in terms of the Pleiades. From there our method for analyzing the Inca
sidereal lunar calendar is to reconstruct the internal logic of the
system of numbers used in counting it and, on the basis of those, to
suggest how the calendar relates to astronomical observation.

So far we do not have any detailed information on time periods
longer than a year. The importance of the calendar as far as we can
retrieve it is strictly in terms of socio-economic and ritual organization and not of historical chronology. Before giving the evidence of the
calendar, let me outline its reconstruction.

Although the period of 328 nights divides into 12 sidereal
months of 27 1/3 days, the Incas did not follow this regular sequence,
but divided the sidereal lunar calendar up progressively into smaller
time units:

1) two equal periods
2) four nearly equal periods
3) twelve periods of rather unequal length
4) 41 very unequal periods.

We can observe a certain general correspondence between the first twofold
division and the two half years of the solar year divided by the
solstices; the second fourfold division and the year divided by the two
solstices and two equinoxes; the 12-fold division and the twelve synodic
months; finally the 41-fold division and a system of 41 weeks of eight
days (41 x 8 = 328) to which certain data seem to point (Zuidema 1977).
Just as the synodic lunar year falls short of the solar year by eleven
unaccounted days, the sidereal lunar year falls 26 days short of the
synodic lunar year and 37 unaccounted days (26 + 11 = 37) in terms of the
solar year.

We can reconstruct with the following known data how the
sidereal lunar calendar was based on astronomical observation. First, it

was correlated to three observations of the Pleiades: their first
heliacal rise in the morning around June 5-8; their first helical set,
around November 18, when the period ended in which the Pleiades were
visible during the whole night; their last heliacal set in the evening
around April 15, after which they remained unseen at night until their
rise on June 5-8. Second, the beginning five of the twelve periods
(referred to as the third division of the sidereal lunar year) were resp.
26, 30, 29, 28 and 30 nights long. This means that, with the exception
of the first period, the other four coincided in length with synodic
months, only the fourth period being one night short. These periods with
a fixed position in the year confirm the special interest of the Incas in
the moon and in synodic months during this time of irrigation and
planting that went from June 9 to October 30. Then they calculated the
deviation of a real synodic lunar sequence in a given year from an ideal
synodic lunar calendar that, because of its fixed position in the year I
will call the fixed synodic lunar calendar. After the initial 5 periods
the close equation between sidereal lunar and synodic lunar months was
rapidly lost. Because of the equation of the second to fifth sidereal
periods to fixed synodic months, we can suggest that the four days preceding the first sidereal period was related to new moon and to the first
days of waxing moon. These were the days of June 5-8 when the Pleiades
would have their first heliacal rise just before sunrise. Eleven synodic
months (11 x 29½ = 324½) are 3½ days short of the sidereal lunar year.
Correlating a fixed new moon with the first heliacal rise of the
Pleiades, means that a new moon will also fall in the last 3½ day period.
Moreover, a full moon will coincide with the passage of the first sidereal lunar half year to the second (5½ x 29½ = 162; ½ x 328 = 164).
Finally, the last heliacal set of the Pleiades occurs around April 15 and
a new moon in the fixed synodic lunar calendar some 10 days after it.
The period of visibility of the Pleiades coincides roughly with that of
the sidereal lunar year.

As a matter of astronomical observation we can say that new
moon rises and sets close to the position of sunrise and -set. Even if
this lunar event cannot be observed, it can easily be reconstructed.
During June 5-8 the sun and the fixed new moon rise close to the Pleiades
rising point, $\pm$ 24° North of East. The Incas observed this coincidence;
moreover, they established a moon-Pleiades coincidence in April. We can
conclude therefore that the sidereal lunar year, consisting of twelve

unequal periods was tied to the time that the Pleiades are visible and to
the length of eleven equal synodic months. In terms of these correlations it would have been practical to start the counting of the sidereal
lunar year on June 5 at the beginning of invisible moon in the fixed
calendar and end it on April 29 just after another invisible moon. The
Incas started it on June 9, four days later. One obvious reason for this
is that they wanted to start their counting after the first heliacal rise
of the Pleiades as well as after the reappearance of the moon in the
fixed synodic lunar calendar. Additional reasons were based on other
astronomical observations correlating the moon to the sun. These reasons
do not affect, however, the scenario of observing the moon and the
Pleiades as given here and its general influence on the sidereal lunar
calendar.

THE EVIDENCE

The astronomical evidence for the sidereal lunar calendar is
based on an identification of huacas in the ceque system in which I have
been employed since 1973; and on a verification, together with A. F.
Aveni since 1976, of the statements in the listing of the ceque system
concerning the importance of certain of its huacas for astronomical
observation on the horizon of Cuzco. The numerical evidence for the
reconstruction of this calendar is based on an interpretation of the
ceque system as a giant quipu used for counting. Data on Inca mythology
and rituals allow us to bring the two types of evidence together and to
confirm the correctness of our reconstruction. I will take the matters
in this order. The astronomical evidence has been published, or is in
the process of publication, by Aveni (1981), Zuidema (1977, 1979, 1981a,
1982a, 1982b), Zuidema and Urton (1976). It will be further analyzed and
brought together in a joint publication. The conclusions reached in the
articles will be used without further need for verification or proof.

Astronomical Observation

The Incas in Cuzco made the following precise observations of
sunrises and sunsets that concerned: 1) the sun alone; 2) the sun in
relation to the moon; and 3) the sun in relation to the Pleiades and the
moon (Figure 1).

The sun alone was observed for the solstices: sunset of the
December solstice was observed from the central temple of the Sun,
Coricancha; sunset of the June solstice from a temple of the sun,

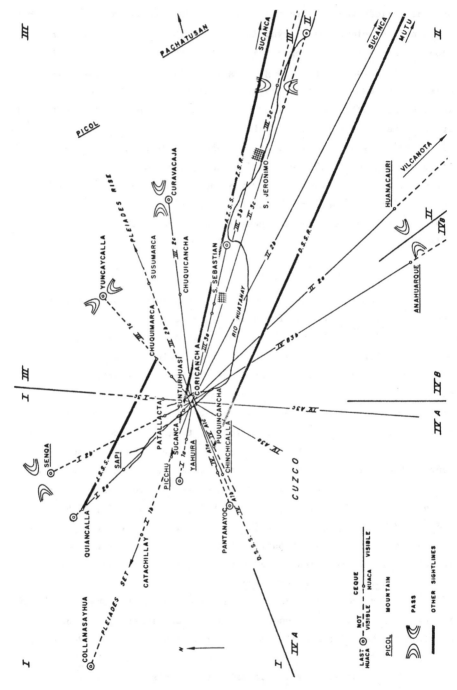

Fig. 1   The ceques used for astronomical measurement

Chuquimarca, North of Cuzco; and sunrise of the December solstice from a temple, Puquincancha, South of Cuzco.

The observations of the sun in relation to the moon concerned first the passages of the sun through Zenith which occur in the latitude of Cuzco on October 30 and February 13, and through Nadir which occur on August 18 and April 26. The midnight sun in Nadir cannot be observed, but the rising and setting azimuths can. On the two days of Zenith passage, when a gnomon will not cast a shadow at noon, the Incas observed sunrise from the mountain Picchu bordering the town on the West. In this way they fixed the days of October 30 and February 13 in the solar year. Taking up a position near the sunrise point on these days, they were able to define the days of anti-Zenith sunset, when the sun goes through Nadir, by observing back to Picchu from that position. These days are half a year apart from the first two, thus the Incas arrived at the dates of August 18 and April 26. Correlating the moon to these four observations of the sun might have been induced already by the fact that when the sun goes through Zenith, the nearest full moon will be close to Nadir, and when the sun goes through Nadir the nearest full moon will be close to Zenith. At least one modern testimony of indigenous culture supports the existence of the practice in observing the moon's passage through Zenith (Meza Bueno 1943). The gnomon can be used for this purpose when the passage occurs during full moon. A disadvantage of this observation is that full moon normally will not occur on exactly August 18 or April 26. Moreover, the moon may deviate from the ecliptic by 5°. Probably in part to circumvent these problems, the Incas not only made an observation of sunset of August 18 from a precise point on the plaza by placing two pillars on Picchu, framing the setting sun, but also by adding two other pillars on Picchu to the right (north) and left (south) of the central pillars that indicated sunset 15 days before August 18, that is August 3 or 4, and 15 days after August 18, that is September 2. Statements of the chroniclers are consistent with a spacing that is equivalent to 15 days of solar motion at horizon. In this way the Incas defined a month of the fixed synodic lunar calendar.

The pillars could be used to demarcate four different months in one year: twice when a full moon or a new moon would occur while the sun set in between the outer pillars and twice when a full moon or new moon itself would set in between the outer pillars. Because the Incas were interested in four dates (two Zenith and two Nadir passages of the

sun) they were also interested in correlating them to the moon at four
times. By chance the four dates of these Zenith and Nadir passages in
Cuzco happen to be times in the following monthly periods: 2½ months
from August 18 to October 30 (74 days); 3½ months from October 31 to
February 10 (103 days); 2½ months from February 11 to April 26 (74 days);
and then 114 days back to August 18. If a full moon falls on the night
of August 18, Nadir passage, then a new moon falls on October 30, a full
moon on February 10 and a new moon on April 26. Only the full moon of
February 10 falls 2 nights short of the Zenith passage of February 13.
Although this correlation is accidental, apparently it was observed by
the Incas since an anonymous chronicler does refer to it (Anonymous 1908
p. 156, 159).

The other observation of the sun in relation to the moon concerns the first possible day of visible new moon of the month that still
reaches the June solstice. According to Molina (1943), the year started
with the first new moon after the middle of May, "a day more or less".
This day in the Julian calendar corresponds to May 26 on our Gregorian
calendar. The alignment of the temple of the Sun, Coricancha, agrees to
such a date as it is not facing the June solstice but sunrise on about
May 25-26 (Zuidema 1982a). A period of 27 or 28 days of visible moon
beginning on May 26 ends on June 21 or 22, the June solstice.

The observation of the rising and setting points of the
Pleiades makes use of one rising point of the sun for one date and a
setting point of the sun for another date. Both of these points also
involve the moon. Moreover the astronomical value of ceque directions
now also becomes clear.

The direction faced by the temple of Coricancha towards
sunrise on May 26 coincides with a ceque that includes a huaca called
Susurpuquio, "the spring of Susur". This spring, already beyond the
horizon of Cuzco, is discussed in a myth about the sungod rising out of
the spring. A poem in Quechua, the Inca language, from the early XVIIth
century relates the name susur to the Pleiades and to the Virgin Mary,
while the latter herself is also compared to the Pleiades in the poem
(Zuidema 1982a). The direction of Coricancha and of the ceque not only
indicate sunrise on May 26, but also are within 1° from the rising point
of the Pleiades in 1500 A.D. Thus, we can understand why the Quechua
poet in 1600 compared the Virgin Mary, as the mother of God in Christian
religion, to the Pleiades, the mother of the stars, (Cobo book 13, ch. 6)

in Inca religion. The myth of Susurpuquio apparently referred to the birth of the Sungod from the Pleiades; stars that in Quechua also are called the "swimming ones", referring to a context of water (Zuidema and Urton 1976). While in Cuzco the first new moon after May 26 was mentioned as belonging to the first month of the year, in other parts of Peru the first heliacal rise of the Pleiades was used for the beginning of the year. New moon in the fixed synodic lunar calendar was before June 9; an approximate date when the Pleiades would reappear. The data from Cuzco and from other parts of Peru do support and confirm each other. Moreover, we now have evidence how the solar, lunar and stellar cycles were integrated to each other in this instance (Zuidema 1982b).

The setting point of the sun on April 26 also was used for observing the setting of the Pleiades around April 15.   Although a similar intent was made as in the first case, the implementation had to be somewhat different, as the sun at that time sets $13\frac{1}{2}^{0}$ North of West, but the Pleiades about 24^{0} North of West (in 1500 A.D.). Sunset on April 26 in between the central pillars of Picchu was observed on the plaza from a place called <u>Ushnu</u> (Zuidema 1979). The pillars themselves, however, were a huaca, called <u>sucanca</u> because of their astronomical function, on a ceque that included also, as another huaca, just beyond the horizon, a spring called <u>Catachillay</u>, one of the names of the Pleiades. The Pleiades set some $2\frac{1}{2}^{0}$ South of the central pillars but still within the outer pillars as seen from Coricancha, that is $2\frac{1}{2}^{0}$ South of the ceque to which the name Catachillay associates them. The date of April 26, as given by the observation of the sun and by the moon in the fixed synodic lunar calendar indicates that the Incas had an interest in the Pleiades also in terms of the date of their last heliacal set.

Some methodological issues of Inca astronomy and calendar

Having given these data of Incaic astronomical observations needed for constructing their sidereal lunar calendar, we can now discuss some of the methods used by the Incas in their astronomy and calendar. The first issue deals with how the Incas used the sun, the moon and the Pleiades in combination to obtain a precise and stable annual calendar. With this purpose in mind, each of the three bodies had for them certain advantages and disadvantages that are of a different nature for each. The sun moves along the horizon either too slow during the solstices or too fast during the equinoxes to be used most profitably for defining a specific day in the year. The moon has a short and rather precise cycle

that can be sharply observed either in terms of first or last visible
moon or in terms of its position against the stars. But its cycle does
not match that of the sun and its movement along the ecliptic has irregularities in terms of the annual cycle. Stars are rather constant over
the years in terms of their rising and setting points and of their dates
of first heliacal rise and last heliacal set. Their disadvantage is in
the observation of these dates exactly. They may be rather difficult to
define from observation, like in the case of the Pleiades. Moreover, the
distribution of stars in heaven is uneven and does not lend itself immediately to a regular sequence of time units by way of first heliacal
rises and last heliacal sets, using a zodiac or any other order of stars.

The Incas reached precision probably in the following order:
First they observed sunset when the sun moves along the horizon at an
intermediate speed, like on August 18. Then they defined a month length
period around this date, like the period of August 4 to September 2.
This period can be defined accurately by observing the possible extremes
of the months whose first or last visible moon still include the day
when the sun sets in between the central pillars of Picchu. The Incas
could obtain this result by counting 15 days from August 18 forward and
15 days backward and observe sunset on those days. From August 18 the
Incas could also count backward to define a specific day for the June
solstice, for first heliacal rise of the Pleiades and for the date of May
26. The Incas used August 18 when the sun is in the anti-Zenith. This
had obvious advantages to them. First, it was related to another astronomical observation, that of the sun's passage through Zenith, that can
be defined sharply in these terms too; second, with a full moon on August
18 there happens to be also a full moon on the June solstice; and, third,
there is new moon at first heliacal rise of the Pleiades; fourth the
periods from the sun's passages through anti-Zenith, Zenith, Zenith and
anti-Zenith happen to be meaningful in terms of the synodic lunar calendar. But the second and fourth advantages are an accident of the latitude
of Cuzco and the third an accident of the precession of the stars fixing a
date around 1500 on which our information is based. The same correlation
of the fixed synodic lunar calendar to the Zenith- and anti-Zenith passages of the sun could not have been made at very different latitudes in
the Andes. August 4, of special importance to the Incas as the official
date on which the Inca king himself opened the agricultural season, is
preserved today for a similar purpose, although now the date is moved to

August 1 as an adjustment to the Western calendar (Zuidema 1981a). But
the date has equal importance all over the Andes from Southern Ecuador to
Northern Chile and it does not seem probable that the Incas from Cuzco
introduced and imposed it. At the moment it seems most reasonable to
assume that the Incas, because of their own ritual concern, wanted their
first synodic month to include the June solstice. When a full moon falls
on the June solstice, there is a full moon on August 18, by chance the
date of anti-Zenith, and a new moon on August 4, close to the date of
exact mid season, August 5 or 6. An advantage of having first heliacal
rise of the Pleiades on a new moon in the fixed synodic lunar calendar is
that then their reappearance can be observed best. While the sun moves
at irregular speed along the horizon and while the cycles of the moon do
not match those of the sun, the Pleiades, as stars, move at regular speed
through the night and through the year. For them equal movement in space
means equal movement in time. Ethnographic examples from outside Peru
demonstrate how this principle has been used especially in relation to
the Pleiades (e.g., Maass 1924). By translating the fixed soli-lunar
calendar into a sidereal lunar one, the Incas measured calendrical
periods in numbers and days.

The second methodological issue concerns how the ceque system
was used for astronomical observation. We realize that the Incas
decided, not only from where they wanted to observe the sun, but also
where on the horizon they wanted to see the sun rise or set for a specific observation. They adjusted two places to each other taking into
account also further considerations that might have no direct bearing on
astronomy. Thus, it resulted that the observations of June solstice
sunrise, June solstice sunset, December solstice sunrise and December
solstice sunset had been made from four different locations. Only the
sunrise and -set observations during Zenith - and anti-Zenith passages
were accomplished within one integrated system. Because of these reasons, the temple of Coricancha could not play a central role for observing the sun. The movement of the moon along the horizon was not measured
directly, but synodic lunar periods were defined by sunrises and -sets.
Pleiades rise was measured by way of a sunrise that happened to be in
about the same direction within 1° and in about the same time (some
12 days earlier). In order to obtain a similar result for Pleiades set,
the setting point had to be observed from two different directions: one
for the sun and the other for the Pleiades. By way of the sun the time

of the last heliacal set was defined sharply for the Pleiades. It was
not necessary for their recognition to define the setting point and
direction of the Pleiades sharply. Neither was it necessary for defining
the time of their last heliacal set. By associating their setting point
to that of the sun within $2\frac{1}{2}^{\circ}$ their timing, however, became well codi-
fied. In only one other documented case do we know how the Incas
intended to relate a star rising point to a particular ceque. This is the
case of α and β Centaurus and of the Southern Cross in the same direction. There may be more such connections. But the ethnographic or ethnohistoric data must be given for such an identification and claim; a
ceque direction by itself was not programmed to specifically define a
star rising or setting. Finally, first heliacal rise and last heliacal
set of any star are not correlated in a regular way to periodicities of
the solar year. While the Pleiades rise was, and could be, attached to
sunrise near June solstice, the ceque towards Pleiades set is in no way
related to sunset near June solstice; even if the ceque could have been
used very well for that purpose, it was not. The methodological problems
that I mentioned, first those of time and then those of space, seem to be
of a comparable nature.

Fig. 2   Schematic representation of the ceque system with
actual borders of suyus. Outer numbers from Cobo;
inner numbers and letters from Zuidema (1964, 1977)

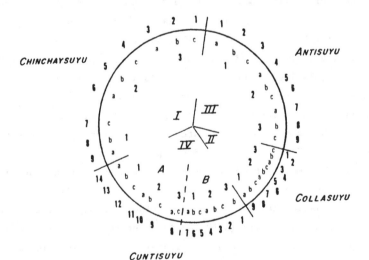

The ceque system as a calendrical quipu.

Although the only description of the ceque system is included
in the chronicle of Bernabe Cobo of 1653 (Cobo 1956; Rowe 1979), its
existence was mentioned by Polo de Ondegardo in 1571 and by Molina in
1573. Matienzo claims that Polo in 1567 learned of all the huacas of
Cuzco from the Indians by their quipus and that they were worshipped in
their specific days (Zuidema 1977). Cobo describes the huacas and their
names, the totality of huacas on each ceque, the rank order of each ceque,
the association of political and social groups to ceques, the totality of
huacas in each suyu (quarter of town) and the totality of huacas of the
whole system. At the end of the list he mentions:

> "the four following (huacas) belong to various ceques but were
> not set down in the order that the rest (were), when the
> investigation was made".

After mentioning these four he says:

> "these were the guacas and general shrines which there were in
> Cuzco and its vicinity within four leagues (+ 20 km   R.T.Z.);
> together with the Temple of Coricancha and the last four which
> are not listed in the ceques, they come to a total of three
> hundred thirty-three, distributed in forty ceques. Adding to
> them the pillars or markers which indicated the months, the
> total reaches the number of three hundred fifty at least. In
> addition, there were many other private (guacas), not
> worshipped by everyone, but by those to whom they
> belonged,..."

My analysis of Inca astronomy and calendar is based on the
clear and unambiguous separation that must be made of the two functions
of the ceque system: one as a system of geographic description,
stressing its spatial function, and the other of it as a quipu, detached
and independent from location, used as a mathematical device for
counting, like a rosary, stressing its function for time, counting days
or nights. Although I am quite aware that the Incas have a single concept of time-space called pacha, we, as outsiders to their culture, can
become aware of their use of space-time only after separating the two
first and integrating them afterwards. Since we know that the Incas used
quipus as calendars and since our evidence indicates that the ceque
system was recorded from such a quipu, we must ask at the moment two
questions: a) what can we say about the text given here by Cobo?; and b)

does the number 328 for the huacas and 41 for the ceques make sense
astronomically and calendrically? Because I shall be concerned primarily
with the second question later, I will answer the first one here.

There is no reason to assume that Cobo did not copy faithfully
the ceque system as it was known already in 1567. But it is also clear,
by verifying on the ground from their descriptions the location of those
huacas whose names are still known, that Cobo and probably the first
recorder of the ceque system did not have first hand knowledge of all the
huacas and of all the directions of the ceques. The huacas were written
down, as Matienzo says they were, from a quipu. The recording was
detached from the terrain. Even though we now know that locations of
huacas were given correctly, the recorder sometimes visualized a huaca
completely erroneously.

Cobo mentions 328 huacas, but then adds to this number four
huacas that were not placed on ceques. Moreover, he adds to the number,
but not to the ceques as directions, pillars that "indicated the months".
Let us analyze the problem of the pillars first. Cobo does not say what
pillars these were and what kind of months they indicated. Elsewhere
(Cobo book XIII ch. 5) he mentions them as being 14 in number and calls
them "sucanca, that indicated the months of the year". He then continues
saying:

> "These were esteemed very much, and they made sacrifices to
> them at the same time as the other huacas and places indicated
> for that purpose. The form of sacrifice to these adoratories
> was that after the sacrifices had been brought to the other
> guacas in the order as the ceques went, as will be said in its
> place, what was left was offered to these markers, because
> they were not in the order of the other adoratories for
> following the order that the sun takes in its path; and each
> one attended with his sacrifice the adoratory that was closest
> to the ceque of which he was in charge". (author's italics)

Cobo probably came to this number 14 (Zuidema 1981a) by taking together
12 pillars sayhua and 2 sucancas; one sucanca for observing sunrise
during Zenith passage and one for sunset during anti-Zenith passage.
Already Polo mentioned both groups, the sayhua and the sucanca, but he
did not add them together. As the research of Aveni and I demonstrates,
the sucancas were part of the pillars indicating months; and at least the
Western sucanca on Picchu was a huaca as part of the ceque system. Cobo

makes the mistake of counting 2 sayhuas twice by not considering them as
sucancas, but the origin of the mistake was already in Polo. We have
identified six rising - or setting points for resp.: 1) sunrise, May 26
as indicated by the alignment of Coricancha and as mentioned by Molina;
2) sunrise, Zenith passage; 3) sunrise, December solstice; 4) sunset,
December solstice; 5) sunset, anti-Zenith passage and 6) sunset, June
solstice. I feel confident concluding that for measuring the sun the
Incas did not use, and did not need to use, any more rising and setting
points than these for defining their calendar. All these data go back,
moreover, to the ceque system, known by Polo and Molina, and to Molina's
analysis of the Inca calendar which can be used in conjunction to and as
a commentary upon the ceque system. The horizon points of 2), 3) and 5)
are explicitly mentioned using the word sucanca; but of these 3) and also
4) hardly could consist of pillars as they are on the distant horizon.
Here the Incas could only have used natural features. Number 5) consisted
of two inner and two outer pillars, and numbers 4) and 6) each of two
pillars or markers. For the measurement of 1) no pillars are mentioned,
although the horizon is nearby. Pillars may not have been needed as the
alignment of Coricancha itself was used. If we assume that in the case
of each 2) and 3) two features on the horizon were used to frame the sun,
then we might suppose that the informants of Polo came to the number 12
sayhuas by adding two sayhuas for 2), 3), 4) and 6) each, four sayhuas
for 5) and none for 1). Although the argument is speculative, it demonstrates that there is no reason to assume that the 12 or 14 "pillars"
were distributed evenly according to 12 months, an argument that I had
suggested before (Zuidema 1964, 1977). As at least nos. 3), 4), 5) and
6) are mentioned as huacas within the ceque system -they were not even
the last huaca on their ceque- we can understand Cobo's text by assuming
that they served a double function: one for the purpose of astronomical
measurement and the other when, in the ceque system as a quipu-calendar,
it was their turn to be worshipped. They were attended twice, on two
different dates and probably each time by different people. Only in the
case of 2) -a rising point that Polo and Molina explicitly mention as
sucanca (Zuidema 1981a)- its location outside the ceque system is possible, although still in the direction of a ceque that ended before
reaching it. The conclusion is that the 12 sayhua or sucanca, when they
served their purpose for astronomical measurement should not be added to
the 328 huacas for purposes of calendrical counting.

The same can be said of the four extra huacas, even though
Cobo says that "they belong to different ceques". We know the purpose of
worship of one. Cobo mentions elsewhere (Cobo book 13 ch. 28) that two
huacas were worshipped in the month of July, one where the irrigation
system of Cuzco started and the other where it ended. The first of
these, Tocori, is the second of the four extra huacas. The last of these
four, Quiquijana, a mountain, might be the same as a huaca in the ceque
system, that might have served the purpose for the irrigation system indicating the end of it. Again the point is, that although Tocori and
Quiquijana served these purposes in relation to the agricultural calendar, this was not part of the ceque system as a quipu-calendar.

Having weeded out misunderstandings and ambiguities from
Cobo's text, we can come now to a final assessment. There were 328
huacas in the ceque system used as a quipu for calendrical counting.
When Polo mentions 12 sayhua and also 340 huacas we know that he had
already made some addition in the way that Cobo did later. By
subtracting the sayhuas, we realize that for Polo, probably already in
1567, there were 328 huacas. Accepting this, the description by Cobo of
the worship of the pillars makes, however, also clear that the order in
which the 328 huacas were worshipped--"as went the ceques", in his own
words--was integrated in some way into the other and very different
order and sequence by which the pillars were worshipped. It is the combination and confrontation of the two orders, one that uses spatial concepts for astronomy and the other that uses a sequential order of huacas
for calendrical purposes, that we are studying.

I have discussed already in detail why the number of 41 ceques
should be accepted (Zuidema 1977). Moreover, I mentioned that probably
there existed an Andean week of eight days. The Inca king made use of it
by having his court every new week ruled by another wife (Pizarro 1978 p.
47, 66) or sister (Cobo book 12 ch. 37). He possibly had 41 wives
(Herrera 1916). These numbers of 328, 41, and 8 (8 x 41 = 328) suggest a
calendrical coordination. We can suggest, moreover, that it is the regular movement of the Pleiades, the mother of the stars, moon and sun,
within the sidereal lunar calendar of 41 weeks of 8 days, that made them
appropriate to measure the other more irregular periodicities that the
Incas observed in the yearly cycle.

Integrating the calendar count into the year

Studying the ceque system as a calendar-quipu, the first

difficulty we are confronted with is the question why it does not account for all the days of the year. In my 1977 article I recognized certain numbers as synodic lunar references (29 or 30) and combinations of others as sidereal lunar ones (e.g. $31 + 24 = 55 \approx 2 \times 27\frac{1}{2}$). I tried to work out a system; a system that may point to certain properties of Incaic calendrical thinking, but that since then remained unverified. I could not take into account a period of 37 unaccounted days ($365 - 328 = 37$) as the data on the Pleiades, from the field and from ethnohistorical sources, were not available at that time. In two recent articles I analyzed these data: in the first article (Zuidema 1982a) on their observation and in the second on their period of invisibility (Zuidema 1982b). The time of last heliacal set for the Pleiades, initiating a period of about 50 days of invisibility, was close to that of the passage of the sun through anti-Zenith, that of new moon in the fixed synodic lunar calendar, and that of harvest (considered in Cuzco and elsewhere in Peru as a time of death). The Earth mother gave up her crop, was conquered by man who harvested, and died, like the Pleiades as the mother of stars. Because the feast of harvest was celebrated during full moon of a movable calendar, Andean culture had no difficulty adapting it to the Christian calendar with the feast of Easter. Even more, the feast of the first heliacal rise of the Pleiades, when they gave birth to the young sun and to the first full moon of the synodic lunar calendar, became easily adapted to the Christian feast of Corpus Christi, 60 days after Easter and thus calculated by and related to full moon. The second article on the Pleiades demonstrated how this period from mid April to early June received its own special ritual attention, making clear that it was a period of death separated from the rest of the year.

Accepting that the ceque system as a quipu-calendar accounts for the period of visibility of the Pleiades, the question becomes: where in the ceque system do we place the gap corresponding to the 37 days, and so where do we start counting and in which direction? The solution that I am going to suggest for a particular reading is based on a comparison of the ceque system as a hierarchical system with the life cycles of the Sun and the Pleiades.

The ceques, organized in groups of three, are ranked as a) **Collana** "principal"; b) **Payan** "second" and c) **Cayao** "beginning" (Zuidema 1964 p. 165). In terms of general sequence of ceques we notice a descending order from West to East, going clockwise through North in I

and III belonging to the Upper Moiety of Cuzco, and going counter clockwise through South in IV and II of the Lower Moiety. In the upper
moiety, Hanan Cuzco, I is higher ranked that III, following the same
hierarchical order as the ceques, but in the lower moiety, Hurin Cuzco,
II is higher ranked than IV, going against the order of the ceques.
There are other features in Hurin Cuzco demonstrating such inversions. I
will mention here the most conspicuous one, pertaining to IV. This suyu
has various irregular features with respect to the other suyus. While
these have each three groups of three ceques, IV has elaborated upon such
an original system creating two groups, called IV_A and IV_B, of seven
ceques (3 + 3 + 1) each (Figures 2 and 3, Zuidema 1977). In the other
suyus there is a descending order of numbers of huacas--with one small
exception--resp. belonging to the 1st, 2nd and 3rd group of three ceques
in each (33, 29, 23 in I; 31, 24, 23 in III; 29, 30, 26 in II) that
allows us to make distinction between "higher" numbers (33,29; 31; 29,30)
and "lower" numbers (23; 24,23; 26). In IV such a descending order is
reversed. By regrouping the numbers here in the combinations (15 + 13),
(15 + 15) and (11 + 11) we come to a reversed order of 28, 30, 22. The
reduplication from 3 x 3 to 2 x 7 ceques leads, in the numbers of the
huacas, back to three numbers but in a reversed descending order.

If now we look at the life cycle of the Pleiades and the Sun,
we notice in both an ascent succeeded by a descent. After the first
heliacal rise of the Pleiades each night they are visible longer, rising
earlier in the night before sunrise. In the nights around their upper
culmination during midnight, they are visible during the whole night.
But after about November 18 during each following night they are visible
for shorter periods of time appearing in the sky already at sunset. If
this happens with the Pleiades during the night, something similar happens with the sun during the day. Although the days become longer from
June solstice to December solstice and then shorter, Andean peoples were
more interested in the fact that during the first half year the sun becomes stronger, rains increase and crops are growing. In the second half
year the sun becomes weaker, rains decrease and plants, especially maize,
after setting seed, ripen and become dry. The rays of the sun, in their
increase and decrease, were compared to the crops, and the harvest was
compared to the sun shaving his beard. The life cycle of the sun was
compared to growth within and on the Earth. But whereas the sun shows
this increase and decrease in day time in a way that we call normal, the

Pleiades show an increase and decrease during the night in a way moving back through the night; and, moreover, when the nights become shorter.

Using the ceque system as a calendar-quipu counting clockwise, from the lowest ranked ceque in II towards the highest in IV and then from the highest ranked ceque in I towards the lowest in III, we describe in terms of the names Collana, Payan and Cayao a similar increase and decrease pattern as with the Pleiades and the Sun. That means that in Hurin Cuzco the calendrical direction goes against the hierarchical direction, while in Hanan Cuzco the two go together. In a language difficult to interpret the indigenous chronicler Felipe Guaman Poma de Ayala may refer to the two directions in Hurin Cuzco in his conclusion to one of his descriptions of the months of the year (p. 885(895)):

> "They say that from the month of January the day is long and the night short and from August the day short and the night long and they say that the moon is one degree low in heaven and the sun is at a high degree and that she is woman and wife of the sun and that the sun is represented with a beard like men and that thus they say that the sun wants to shave the hairs and harvest of his beard."

On the previous page, Guaman Poma insists on relating the shortest night to August and not to June, although he does mention here the feast of St John (June 24). His reasons are not astronomical but calendrical, because August was the month measured as a fixed synodic month by the four pillars on Picchu when the sun goes through Nadir, but the moon at night passes close to Zenith. In the next section I will support with specific data an Inca concept that now we can mention in general about the existence of two directions in Hurin Cuzco. The moon is strongest, at night, when the sun is weakest. The moon loses influence when the nights have become short and when the sun has gained strength. The two directions exist in Hurin Cuzco because of the joining that the Incas allow of two concepts of direction: one of hierarchy and one of time. I stated the problem first in terms of the sun and the Pleiades because they represent the principle of increase and decrease in general terms and because the Pleiades are needed for connecting the sidereal-lunar calendar to the solar year. But the principle of two directions in Hurin Cuzco is also apparent in terms of the moon. In the next section I will discuss still another and even more explicit use of this principle in Cuntisuyu (IV); a descending order of ceque names in one direction

counterbalanced by a descending order of numbers of huacas into the other
direction.

THE SIDEREAL LUNAR CALENDAR AS AN INSTRUMENT OF ASTRONOMICAL MEASUREMENT

When we come now to assess the value of the ceque system as a
quipu-calendar and thus as an instrument of astronomical measurement, we
have to ask two questions: first, does it work? and, second, where is
the evidence? Why, apparently, is the evidence so difficult to come by?
The second question is, in final instance, the more interesting and
lengthy one, but can only be answered after one has obtained a good,
although still hypothetical, fit between solar and sidereal lunar year.
We saw that data on the synodic lunar calendar derive from seasonal activities. Having solved the problem of the month of sowing around the
observation of the sun on August 18, the other data are rather clearcut
and unambiguous. The data on the sidereal lunar calendar derive,
however, completely from political organization; data that from a Western
point-of-view do not seem to be of an immediate calendrical character.
Here our approach depends on taking up a good strategic position in analyzing data on socio-political organization, kinship, dynastic theory,
professional groups, myth and ritual, and on having at least one
beginning clue of relating one such group to a calendrical event. In the
next section I will discuss one such a clue that confirms the analysis
of the calendar as will be given here. In general terms we cannot say,
however, that the chroniclers do not support strongly the position about
the ceque system as a calendar. The anonymous chronicler who gave us
also the best information about the pillars on mountain Picchu and their
significance for observing the sun in August, says (Anonymous 1908 p.
150):

> "The next Ynga (king), called Ynga Yupangue, put more order in
> Cuzco, as head of his kingdom and court. He ordered and
> divided the (proper) Ynga (population) into 12 social groups
> ("parcialidades"), so that each group would account for its
> own month, adopting for itself the surname and name of that
> lunar month; and (that group) was obliged on the day that
> began its month, to go out to the plaza to announce its month,
> playing trumpets and shouting and howling, so that it was
> manifest to all. In this order all the months of the year

were very well organized and they celebrated them with many
ceremonies and sacrifices that they did to the Sun and to
their guacas and ydols, with their priests that for that purpose they had of indians, witches and impostors. And he
ordered that in his whole kingdom, in each province this order
was kept by the Governor who was in charge of each province,
because they were already well instructed as something well
known to them".

The chronicler makes clear how much we are dealing with a state calendar,
a political instrument. But still he talks about (synodic) lunar months,
even if then he continues to describe its complications in terms of the
observation of the sun in August; and even if other chroniclers like Polo
and Guaman Poma (Zuidema 1977) are aware that these months could not be
just synodic, because, as they say, the number of days that the lunar
year was short on the solar year was distributed throughout the year and
the days were added to different individual months. The social divisions
of the Incas proper –all descendants from the first king– correspond to
ten panacas or "royal ayllus" that Cobo describes, each in relation to one
group of three ceques. In that way he confirms the calendrical context.
But he, and other later chroniclers confound the issue by claiming that
each of the first ten kings of the dynasty had founded his own panaca,
composed of his descendants with the exception of the crownprince and his
sister-wife. The problem is far more complicated than thus described and
the calendar and other data will help us to give a far different picture
of the panacas as groups distinguished by origin, rank, profession
and by moiety and suyu. There were ten panacas of the royal Incas in
Cuzco and two groups that represented the non-Inca population, not
descending from any Inca king in the male line. While these data on
social structure mention the sun and the moon as calendrical determinants,
data from other parts in Peru mention the stars, or the moon in conjunction to the stars. The mythology of Huarochiri in the XVIth century, in
Central Peru, refers to stars in this respect (Avila ch. 29; Zuidema and
Urton 1976). The chapter on stars finishes by saying:

"Before only a part of the people worshipped these stars as
powers who gave life and who gave shape. Other people dedicated themselves in the hope of adding to their strength and
power. They gave cult to these when they came up, waking the
whole night".

The text discusses here the upper culmination at midnight of stars and
constellations. There is a distinction between two types of simultaneous
cults to the stars: one because of reasons of descent from the stars and
the other because of reasons related to private or professional interest.
The chapter gives an example of how especially llama owners worshipped
the constellation of the llama. At the moment our interest is in
noticing that only stars and not the sun or the moon are mentioned,
although we must assume that in this part of Peru with a similar calendar
as in Cuzco we are dealing with comparable political interests in the
calendar. The particular example of a myth from Cuzco, to be discussed
in the next section, will show that the calendar interest of a social
group was influenced by the sun, moon and stars together. Finally, a
myth from the North coast of Peru (Calancha 1639 pp. 552-554) mentions
explicitly how in one case the moon was worshipped in conjunction with
the Belt of Orion. The general data mentioned here may give us confidence in our endeavor of formulating a theory for a sidereal lunar
calendar. But they should warn us also that, where a complex society
uses complex astronomical data for its political needs, we should expect
also a complex solution.

The numerical data

The ceque system as a calendar tried to adjust, accomodate and
integrate many interests, making an advantageous use of various astronomical concurrences and coincidences. Here I will suggest a specific
reading of the ceque system as a calendar, to be supported with textual
evidence in the next section. Although my intention is to go from the
more general problem to the more specific issues, they cannot be
separated altogether from the beginning. In Figure 3 each day is given
the space of one degree. The calendar starts on June 9, the day of
heliacal rise of the Pleiades, and reads clockwise. The space in between
two radii represents the number of huacas on one ceque. The ceques are
counted hierarchically from I 1 a through sections I and III and from IV_A
1 a through sections IV and II. This means that in the calendrical use
of the ceque system a space belongs not to the radius that follows it in II
and IV, but to the radius that precedes it in I and III. Radius I 1 a
therefore coincides with IV_A 1 a. The first ceque of a group of three is
represented by a longer radius, called a, and is followed by the ceques
and radii b and c. The names in the center belong to the synodic months
in the fixed lunar calendar with their suggested correspondence to the

Fig. 3   The ceque system as a quipu-calendar

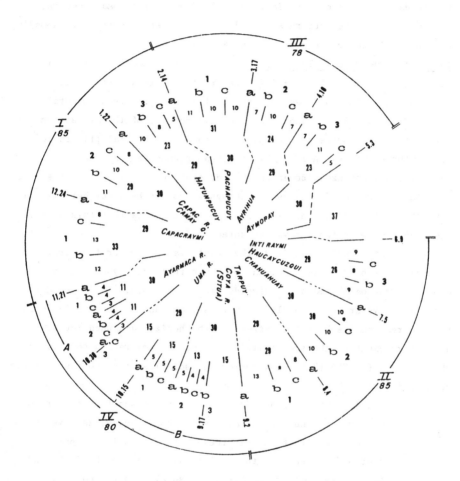

sidereal time units. The dates belong to the radii, that is to the day
after the night represented by the space of one degree in II and IV and
to the day before the night in I and III. When using roman numbers, ara-
bic numbers and the letters a, b and c, I will refer to the corresponding
time units in the ceque system as a calendar.

The first five synodic lunar months. In the case of II 3, II 2, II 1,
(IV_B 3 + IV_B 2) and (IV_B 1 + IV_A 3) there is an almost perfect fit with
the fixed synodic months as measured by the four pillars on mountain
Picchu. II 2 a, II 2 b and II 2 c measure periods of 10 days as the sun
advances to the Northern outer pillar through the central pillars to
the Southern pillar. Avila (ch. 18), Guaman Poma (pp. 235(237),
260(262)) and Betanzos (Betanzos ch. 18) mention how synodic months were
divided into weeks of 10 days. Although we do not have further textual
support for this statement, the ceque system suggests that this division
might be true when the sun was measured in July.

       With 26 days in II 3, the count started, after the first
heliacal rise of the Pleiades, and when, in the fixed synodic lunar
calendar, the waxing moon was already clearly visible. II 3 b measured
the eight days around June 21-22, that is from June 18 to June 26, when
the sun reached its Northern extreme on the horizon and when for some
days no movement of the sun is visible. Guaman Poma refers to this fact
when he says (p. 884 (898)) how during the solstices the sun rests in his
seats and that:

> "so (is) the turning of the wheel of the sun from summer to
> winter from the month that begins from January, as says the
> philosopher, that one day the sun sits down in his principal
> seat and lordship in that principal degree and he rules and
> takes possession from there, and so also in the month of
> August, the day of St John the Baptist, he sits down in the
> other seat, and so coming from the first seat to the second
> seat he does not move from that seat in this most important
> day, from there he rests and rules and governs as a king in
> that degree; the third day he moves and prepares himself for
> his voyage, only one minute, very little, as it is said that
> he prepares himself for the travel and from that degree he
> travels (walks) each day without rest, for about half an hour
> towards the left, facing the North Sea (= Atlantic Ocean) for
> six months from..." January to July and then "...from August

> he begins again from the principal seat to the second prin-
> cipal as he has two seats much in his power as each month has
> its own seat in each degree of heaven (where resides) the sun
> and the moon follows (him) as (his) wife and queen of the
> stars she follows the man who indicates the clock of the
> months of the year..." from August to December.

If for a moment we forget about the allusions to the August turning point
and read how the second seat is equated to the day of St John (June 24),
we realize Guaman Poma's understanding of the sun not moving during the
solstices, a period that the ceque system counts as eight days, not only
during June solstice, but also as we will see during the December
solstice.

One ritual of the month of June, more than anything else,
exemplifies how ritually the ceque system became such a powerful instrument for counting time. Moreover, it may be a support for the use of 26
huacas and days in this month. Molina (pp. 26-27) describes how in the
month of **Inti** **raymi** "the feast of the Sun" - when during the whole month
the Inca king himself went to the temple of **Chuquimarca,** from where, as I
concluded, he observed June solstice sunset- priests went to the temple
of **Vilcanota,** "the house of (the birth of) the Sun", on the continental
divide between Cuzco and Lake Titicaca (see Fig 4 ; Zuidema M.S.a.). At
the moment it is of no concern why they considered the direction of such
cosmological importance, nor why they started the pilgrimage at the sucanca for December solstice sunrise and why they ended it at the sucanca
of the sunrise during Zenith passage. There are data that the same pilgrimage was carried out from October 30, Zenith passage, to December solstice. Of interest at the moment is that Molina describes how during
this pilgrimage:

> "(the priests) went to burn llamas and other offerings in each
> day of this month in the following places...".

Of the 21 places that Molina mentions, I have been able to identify at
least 13. If the informant of Molina intended that only on 21 days
offerings were made, then the statement is not true that during all the
days of the month offerings were given. But the distribution of the
known huacas on the roads to and from Vilcanota makes clear that those
mentioned after 8 and before 14 (these two numbers are not shown on Fig.
4) were visited twice. In this stretch, the pilgrimage followed the
river in both directions, while before on the way down it went through

Fig. 4   The pilgrimage route to and from Vilcanota

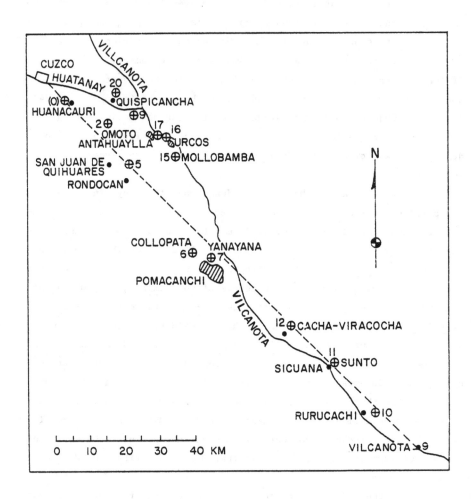

the mountains, and after, on the way back, it continued following the
river. If five huacas were visited twice, then during 26 days offerings
were made, a number equal to that of the huacas in the ceque system of II
3: local huacas that were attended during 26 days by people when at the
same time priests went much further away. While I recognize the hypothetical character of reconstructing the number of days involved, the
discussion indicates a direction of research on important material that
clearly represents an Andean way of handling problems.

  If now we turn to the last month of the 5 month period, the
end of this month is marked by the passage of the sun through Zenith on
October 30. With a full moon on August 18, at that time a new moon will
occur; a fact to which, as I observed before, the anonymous chronicler
refers. After October 30 an extra period of 22 days remains until the
beginning of the second sidereal lunar half year. Before analyzing
further the months of the five month period and this period as such, let
me discuss the importance of the date of November 21.

The date of November 21. On November 18 the Pleiades have their first
heliacal set before the sun rises; that is, they are last visible in the
West in the pre-dawn twilight. Some three days before, full moon
occurred in the fixed synodic lunar calendar. While at the time of first
heliacal rise of the Pleiades new moon in the fixed lunar calendar rose
after the Pleiades, now on November 18 the moon may have overhauled them.
These are two reasons why November 21 could have been chosen as the
turning point in the sidereal lunar calendar: the first being that
after the periods of heliacal rise and upper culmination, the Pleiades
now begin their other two periods of heliacal set and invisibility; and
the second, their relationship to the moon. But both arguments cannot
explain why the Incas preferred November 21 finishing a period of 165
nights, one night longer than half a sidereal lunar year.

  Three observations may help to solve the problem. First, the
period of I 1, with its 33 huacas and nights, ends on December 24, just
after the December solstice. We have ample evidence that the ritual purpose of this month was to end with the latter event. We can suspect
therefore that the choice of November 21 has to do with a reasoning
related to the solstice. Second, the sidereal lunar half year represented by the 165 huacas of II and IV is 12 nights short of six synodic
months; 12 nights that may have been represented by the 12 huacas of
ceque I 1 a. Data related to I 1 b and I 1 c that I will mention in a

moment confirm this suggestion. Third, the excess of 12 days is close to the difference between synodic lunar half year and sidereal lunar half year (177 - 164 = 13), but also to the difference between solar year and synodic lunar year 365¼ - 354 = 11¼). If we can accept the 12-day period of I 1 a as a compromise between the two periods of 13 and of 11¼, then we have here an additional support why the temple of the Sun, Coricancha, was aligned towards sunrise on May 26, as measured by Aveni and me and as supported by the chronicler Molina when he says that the year began with the first new moon after May 26. Having an ideal, fixed full moon on the June solstice would allow the Incas to predict that the month of Inti raymi in the next year would still begin just after the observation of sunrise on May 26. Coricancha was aligned as it was because of the difference between the tropical year of the sun beginning after May 26 and the twelve fixed synodic lunar months of the year beginning on June 9. Thus the Incas were also helped in organizing the celebrations around the December solstice as planned according to the periods of the ceque system. We can read in the number 33 for I 1 the sum total of 27-28 days of difference between May 26 and June 21-22 plus the 5½ days that by December 24 the synodic lunar half year (177 days) was short of the solar half year (182½ days). These calculations support the further calendrical evidence to which we come now. While the June solstice, with its small feast of the sun, was associated to only one synodic month and to an ideal full moon fixed at the date of the June solstice itself, the December solstice, with its great feast of the sun, was represented by two months, one before and one after. The next paragraph discusses how this was done.

The celebration of the December solstice rituals. If the rituals of the June solstice suggested already how a close description of these can be used for reconstructing the calendar, this is even more the case with our data on the two months around the December solstice: Capac raymi the "royal feast" and Capac raymi Camay quilla "the royal feast, the moon of animation, creation, production". (See Taylor 1974-76, Duviols 1978 on camay). Before (Zuidema & Urton 1976) we observed that Capac raymi is the only month without a lunar name, and that the second month combines the first name with a lunar name that does not refer directly to an agricultural or seasonal activity. Only of these months do we have an account of the day to day activities, dedicated to the initiation rituals of the boys of nobility. While the rituals in the first month in no way refer to

the moon and end on the December solstice, the rituals in the second month
only refer to days in relation to phases of the moon. Capac raymi, then,
appears to have been a month calculated only in terms of a fixed calendar,
but Camay quilla only in terms of a movable calendar. Let us look more
closely at the rituals of these two months and see to what extent they
correspond to the data of the ceque system.

Molina describes for Capac raymi:

1st - 8th days preparations for initiation of the boys;

9th day their hair is cut; they sleep at foot of mountain Huanacauri;

10th day they go up to Huanacauri;

11th - 13th days rest;

14th day they leave the plaza; they go and sleep at the foot of mountain
        Anahuarque;

15th day they go up and race down Anahuarque;

16th day they receive earspools and come back to town;

16th - 21st days other people execute a dance with puma skins on the
        plaza;

21st day the boys bathe and receive new dresses;

22nd day their ears are pierced as the last and most important ritual
        in making them men.

23rd day Visit by the Sun of the June solstice to the temple for obser-
        ving sunrise on the December solstice.

If we integrate these data and days into the ceque system, my suggestion
would be that the first part of the feast, including the visit to
Huanacauri and the three rest days after, correspond to I 1 b, consisting
of 13 huacas, and that with the visit to Anahuarque the last part of the
initiation and the counting of I 1 c begins. This would mean that the
inititation terminates with the ritual bathing on the 21st day, when
Capac raymi also would end. The new dresses, the piercing of the ears
and the visit to the December solstice temple already would belong to the
next month. The number of days of the ritual is very close to that of I
1 b and I 1 c in the ceque system. I 1 c would correspond to a similar
period of eight days around the December solstice as II 3 b for the June
solstice.

Molina's description of Capac raymi Camay quilla is as
follows:

1st day of moon recently initiated boys have a ritual battle between the
        two moieties; no fasting anymore;

1st and 2nd days the dance Yaguayra;

3rd to 14th days work in the cultivated fields;

15th day return to Cuzco on full moon; dance Yaguayra

16th and 17th days Dance with rope Muruurcu

18th Pray for rain in dresses Angas onco "blue dress" and Quillapi onco
    "moon dress".

18th to 20th day the dance chupayguallu (chupa "tail")

19th day they follow sacrifices from the place in Cuzco called Pumap
        chupan "tail of the puma" down the river to Ollantaytambo;

20th day recently initiated boys run back from Ollantaytambo in com-
        petion;

21st day and onward the days that remained of this month were dedicated
        to work by people in their own fields.

We see that the important rituals were carried out from full moon on the
15th day to the 20th day. The last, probably eight, days of visible moon
were dedicated to people's own work. Ritual attention to the moon in
this month corresponds to the 29 nights in I 2. Thus the last 8 days of
no ritual could correspond to I 2 c. We observe, however that I 2 and
the corresponding fixed lunar month do not overlap completely. Probably
we have here one reason why the first half of Camay quilla was not important ritually. With a full moon of the fixed lunar calendar on January
12 – as there is a new moon on October 30, and thus there are full moons
on November 14, December 13-14 and January 12 – the earliest possible full
moon would fall on December 28. This date would assure that the rituals
of Camay quilla did not interfere with finishing those of Capac raymi as
these were celebrated only before and including the December solstice.

        The fact that Capac raymi was such a long month appears to
have been known by Guaman Poma. When he concludes his first discussion
of the Incaic months, he says (Guaman Poma p. 260 (262)):

        "in all the months and years they added these months and days.
        From one to ten days is a week , and so they came to 30 days
        or 31 days or 32 according to the waning moon, all that as
        they had ordered from waxing moon the 12 months that were
        counted in a year and in this order they made their quipus of
        expenses and multiplications about all that occurred in this
        kingdom each year".

The data on Capac raymi, I 1, and on only one other period, that of III
1, do confirm Guaman Poma's statement. It seems most likely that his

statement pertains to these facts, as otherwise it would not make any sense. The discussion on Capac raymi clarifies the statement by Polo and others (Zuidema 1981a) of an original Inca calendar beginning on the December solstice but later moved back to one month earlier. The date of November 21, initiating the second half of the sidereal lunar year, began the fixed month of **Capac raymi** "the royal feast". It was the month of **Capac ayllu**, the "royal clan" that as a panaca belonged to I 1; and, whereas all other ceques of the ceque system were called either Collana, Payan or Cayao, the name of ceque I 1 a was **Capac**, "royal". It was the highest ranked ceque. The data on Capac raymi and Capac raymi Camay quilla support decisively our reading of the place of the ceque system as a calendar-quipu within the solar year.

The second sidereal lunar half year. As the argument led me to discuss the first two months of the second sidereal lunar half year, and as I prefer to end the argument with a further discussion of the calendrical importance of IV, I will mention the other data on this half year. The period of III 1, of 31 days, began with the second passage of the sun through Zenith, and ended March 17, 4 days short of the equinox; an event that neither here nor in the case of the September equinox was given attention. Chroniclers give no information, so probably the equinoxes were unimportant in Cuzco. Period III 2 ends on April 10, one day before the sun passes through the first Southern pillar on mountain Picchu and about five days before the last heliacal set of the Pleiades. Period III 3 was to end the sidereal lunar year. The most important question about these four periods (from I 3 to III 3 inclusive) is why III 1 received 31 days. I can see no reason, except a socio-political one. Clearly, more important periods and more important panacas were given more days. While I 1 belonged to Capac ayllu, representing the high nobility, III 1 belonged to **Sucsu panaca**. This panaca, according to dynastic mythology, stemmed from the king, **Viracocha Inca**, who derived his name from **Viracocha**, by the Spaniards identified as the Creator god. To his panaca belonged the high priests in Cuzco; associated with it, in III 1, was the ayllu of the other priests. On this basis a possible argument can be given that this panaca, after Capac ayllu, should be assigned most days. In the second sidereal lunar half year the discrepancy with the fixed synodic lunar half year became more marked. After Camay quilla the rains diminished and the crops were ripening. The agricultural concern in terms of the moon became less prominent.

The time periods in IV. We come now to ask the question whether there
existed astronomical and calendrical reasons for the observed singular
position of IV and for the irregularities that were mentioned before.
Why was IV divided first into two sub-suyus IV$_A$ and IV$_B$, and why went the
hierarchical order of the numbers of huacas per ceque against the hierarchical order of the ceques as indicated by the names Collana, Payan
and Cayao? Even more curiously, looking at the numbers of huacas we see
that there is more order involved, not less, than in the other suyus.
Ceque IV$_B$ 3 b -one of the few having an individual name, Anahuarque- and
ceque IV$_A$ 3 a,c -one ceque composed out of two- each have 15 huacas,
equal to one half of a synodic month. IV$_B$ 2 and IV$_B$ 1 have a sequence of
huacas per ceque of 4, 4, 5, 5, 5, 5 and in IV$_A$ 2 and IV$_A$ 1 one of 3, 4,
4, 3, 4, 4. Furthermore, IV$_B$ 2 + IV$_B$ 1 together have 28 nights, close to
a sidereal lunar month of 27 1/3 nights, organized around a new moon on
September 30 in the fixed synodic lunar calendar. This period of 28
nights and the whole period of IV would have been a useful way to measure
the moon against the stars. Many data allow us to study this problem.

Three constellations near the South celestial pole are important today and were in the sixteenth century (Zuidema & Urton 1977; Urton
1981). α and β Centaurus are and were known as Llamapa ñahuin, the "Eyes
of the Llama". From there stretches northward the dark cloud constellation of the Llama, in Huarochiri known as Yacana. Under and near the end
is the baby llama suckled by her mother. In front of the Llama was and
is the dark cloud constellation of Yutu - yutu a tinamou bird, in
Peruvian Spanish called "partridge" -that in Western astronomy is called
the "Coalsack", within the Southern Cross. Probably the Southern Cross
itself was also important, but here the older data are not clear. Today
it is known as Yutu-cruz, the "cross of the Yutu". In Huarochiri the
only starmyth, mentioned as such (although there may have been others)
concerns the Llama Yacana. When during midnight she is invisible she
drinks the water from the Earth preventing the Flood. She makes men who
worship her rich in wool. When she rises, she suckles her baby, who then
still has not risen. Another important myth from Central Peru in the
sixteenth century (Duviols 1974-76 pp. 275-77) discusses how a foster
mother -whose name, Pullucchacua, according to the myth means "skin
without feathers of the partridge"- suckles the Thundergod on a misty
morning after he, during the preceding night, is reborn. The Thundergod
immediately grows to full size and defeats his enemies. The reference

to the Thundergod and the night place the partridge—mother in an astronomical context. The theme of her suckling on a misty morning, when
sunlight is still dubious, is comparable to the Llama suckling her baby
when she has risen already but the baby not. Therefore, we can recognize
in the partridge—mother the dark cloud constellation of the Yutu when she
rises just before sunrise and in the suckling Llama mother the Yacana
constellation with her appearance also just before sunrise. Llamas and
partridges were the most important sacrificial animals. One early source
from 1539 (Anonymous 1879) reports how llamas and partridges were sacrificed on mountaintops during new moon. Black llamas were sacrificed by
starving them in October in order to make them weep, sympathetically
calling for rain. The description of Yacana in Huarochiri clearly refers
to the lower culmination of this constellation. For these reasons we may
conclude that the constellation of the Black Llama as well as that of the
Yutu were important astronomically in terms of their period of lower
culmination and first heliacal rise.

     Notwithstanding the great importance of the Pleiades in the
sixteenth century and now, and of the various names under which they are
known, I am not aware of any important myth discussing them. There are
various old and modern references to their importance for prognostications in agriculture and llama husbandry. Different times of the year
are mentioned when they are used for that purpose (Urton 1981); June,
August and, as was shown by informants to me, November. The most important point of consideration is if they are observed big or small,
brilliant or opaque. Finally, the same term **Catachilla** is used in Aymara
(Bertonio 1956) for the Pleiades, in **catachilla huarahuara** "star
catachilla" and for the Black Llama constellation in **catachilla**
unuchilla. (Unuchilla probably derives from **unu**, the root of **unuqueatha**
"to move", used in **laccampuna unuquenapa** "the movement of the sky".)
But, as we saw, the term catachilla for the Pleiades was also known in
Cuzco, with the possibility that the other connotations that it received
in Aymara were also known and used.

     The stars of the Southern Cross and of α and β Centaurus,
leading the Yutu and the Llama, distinguish themselves because of their
position close to the South celestial pole. At the latitude of Cuzco,
they are never invisible during the whole night. Around the time of their
lower culmination at midnight they can be seen setting just after sunset
and rising just before sunrise. Because of this phenomenon, we can call

them quasi-circumpolar. The text of Avila on Huarochiri probably refers to this. But because of this situation, first heliacal set will occur before last heliacal rise, contrary to what stars will do that are further removed from the South celestial pole. The quasi-circumpolar stars reverse the normal order; the data of the ceque system in IV seem to reflect this situation.

The period of lower culmination for the Southern Cross, extending from its first heliacal set to its last heliacal rise starts (in 1500 A.D.) from about September 3 to about September 29.

The period of lower culmination for α and β Centaurus is from about October 7 to about November 2.

The period of upper culmination for the Pleiades is from about November 5 to about November 18.

The joint observations of these three constellations coincides very well with the period of IV from September 2 to November 21. Especially the period of September 2 to October 30, spanning the two synodic months that the sun moves out of the four pillars on Picchu to the time that he goes through Zenith and the fixed new moon through Nadir, was of great interest to the Incas. Then they could observe the quasi-circumpolar stars in relation to the movement of the moon through the stars. The data on this star group and on the Pleiades suggest, moreover, that their movement was seen not only conjointly but also in opposition to each other. When one had its upper culmination the other was close to its lower culmination and vice versa.

But the question still remains, why were the Incas so interested in the period of September 2 to November 21? Were there any other reasons than the structural ones given here? The discussion in the next section will support the argument in terms of Inca concerns about fertility, cosmology and political organization; it will give written support for the place of the sidereal lunar calendar in the solar year as suggested here. There I can come also to a closer analysis of the data demonstrating that the sidereal lunar month of 27 1/3 nights actually was observed within that period. Here I will approach the question only in terms of general concern with agriculture and rain.

After the Incas had measured the sun behind the four pillars of mountain Picchu, during the month of August, their month of early sowing, and after they had determined thus the correlation to the movable synodic month of the year and its agricultural activities, the months of

intensive concern about sowing, planting and the coming rains were from
September to November, when these concerns became continued by the initiation rituals; rituals of fertility not anymore of plants and animals
only, but also of humans. When at September 2 last correlation of moon
to sun was measured, it may have been replaced by measuring the correlation of moon to stars. From September 2 to October 30 the synodic months
could be measured in their advance through the stars. In the period from
October 30 to November 21 the concern was to come back to the correlation
that the Pleiades had established and started in June 9.

Before finishing this section, I have to come back to the data
of Guaman Poma about the two seats of the sun and his division of the
year into unequal periods: one of seven months from January to July
inclusive and one of five months from August to December inclusive. Even
if the argument to us seems confused, he is consistent about it throughout his chronicle. Elsewhere he says (Guaman Poma 1152 (1162); Zuidema
1981) that the Inca king ritually opened the agricultural season on the
day of Santiago, July 25, and claims that this in fact was the beginning
of August. He was aware that July 25 in the Julian calendar corresponds
to August 4 in the Gregorian. Thus apparently he knew the importance of
the date when in Cuzco the sun went past the first pillar of mountain
Picchu (see also Guaman Poma 884 (898), 885 (899)). The five months of
August to December are linked by him to the moon, probably because of the
agricultural concern; he terminates this period with December as the last
month of planting. But we notice the discrepancy with my conclusions of
the five synodic months -those governed by the moon as Guaman Poma would
say- being those from June 9 (or, with the days of new moon included,
from June 6) to October 30. There is no reason to dismiss Guaman Poma's
information. But if the discrepancy is a genuine one, not belonging to
any misinterpretation of data, then we must conclude that probably it was
already noticed by Guaman Poma himself as he identified the second "seat"
of the Sun as well with Saint John, June 24, as with Santiago (Saint
James), July 25, for him the beginning of August.

MYTH AS EVIDENCE FOR THE CALENDRICAL USE OF THE CEQUE SYSTEM
AND ITS PLACE IN THE YEAR

In the last section I gave a description of what seems to be
the best fit of the place that the ceque system as a quipu-calendar
occupied within the solar year.

If we can accept this general framework of the sidereal lunar calendar as correct, even more worrisome questions arise: why did the Spanish chroniclers not report on it?; why were they not aware of it?; why, apparently, did the Inca informants not report more explicitly on that matter? One reason may be that the numbers we had to deal with on first view look very irregular as they count actual distance in time between unconnected events. The more important reason probably is, as I mentioned before, that the knowledge of these periods was encoded in data on social organization and in myths and rituals related to it and not so much in data of seasonal events of agriculture.

In this section I will argue that the data we are looking for may not be missing. In one case we have precise data on the relationship between a political group and its place in the ceque system; a myth and a ritual function belonging to it, an astronomical observation that the latter comment and a precise date in terms of the Christian calendar that confirms the importance of all this information by tying it to the Incaic sidereal lunar calendar. The myth comes from a late chronicler, Murua, who was proven to have committed major plagiarisms. In a sense this makes him more trustworthy as it can be expected that the data to be discussed here also derive from a good, older source; a source that I suspect may be Polo de Ondegardo. The myth may seem a minor one, but it connects with many others in Inca culture. As it does serve its purpose of anchoring the ceque system as a calendar into the solar year, the argument will allow for expansion with its interconnections with other themes. A full treatment of it is beyond the scope of this article, but the calendrical argument and its implications are clear and can be stated straightforwardly. In my discussion of the myth I have to make use of arguments on Inca history, myth and the dynasty that I have discussed elsewhere more at length (Zuidema 1964, 1973, 1981b).

The myth of Cusi Huanachiri

The myth as given by Murua concerns an Inca hero called **Cusi Huanachiri** or **Manco Inca**. I will give a full analysis in another article (Zuidema, M.S.b.). Here I will only reconstruct the bare outline of it as it relates to the study of the ceque system as a calendar. There are two manuscripts of the chronicle (Murua 1946 and Murúa 1962). The more extensive version of the myth is found in the first manuscript Murua 1946 book 2, ch. 1,2. According to this myth, Cusi Huanachiri is a grandson of the founder of the Inca dynasty, **Manco Capac**. Murua tells the myth in continuation to a myth about Cusi Huanachiri's father **Inca Yupanqui** who, after the event told in his myth, was called **Pachacuti Inca**. This other myth is a version of the story mentioned by other chroniclers of Inca Yupanqui or Pachacuti Inca as the ninth king of the dynasty and not as a secondary son of the first king. Guaman Poma refers to Cusi Huanachiri (Guaman Poma p. 149) with very similar data as Murua, placing him in the same chronological context. He makes him, however, a brother of the third king, **Lloque Yupanqui**, and he does not give the calendrical and hierarchical implications of the myth.

The myth links the following five arguments to each other:
1) Cusi Huanachiri as a warrior invented the custom of ear piercing;
2) he also developed the idea that, whenever one wanted to drink, he should first offer a drink to the Sun;
3) because of the conquests that he and other captains had made, the Incas celebrated on the first of October one of the feasts that was held in the whole kingdom. "And any family and Indians in their huts would make invitations, like they were accustomed to do, and there they made their sacrifices close to the fires to their gods and huacas and specially to the Thundergod and to **Tipsi Viracocha**, whom they held as their Creator of the World."
4) "This prince and captain **Mango Inga** was married with a **ñusta** ("princess") one of his cousins; he had many sons, in her and in others, that people feel sure were more than a hundred; ...; of them people say that they married in this town with some ñustas called **yunacas** (= **iñaca**) who were also principal ladies, and to each of them they gave 150 women as   service, who were of those that this coura- geous prince Mango Inga brought from the war against many other peoples."
5) the other name of **Cusi Huanachiri** was **Manco Inca**.

Each of the themes of this myth is mentioned also elsewhere. Cieza, our first chronicler, who published a full history of the Incas in 1551 and an account of his travels through the Andes, mentions (Cieza 1945 ch. XCVII) that the Incas learned the custom of ear piercing from the **Caviña**, a people living further up along the Vilcanota river. Guaman Poma reports (pp. 100, 101), not only how Cusi Huanachiri had to drink to the Sun in order to give battle to his enemies, but also how the custom was invented by   **Capac Yupanqui**, the fifth king. This data is

important as a similar version is found in the chronicle of Santacruz
Pachacuti Yamqui (pp. 231, 233), linking it to the pilgrimage around the
June solstice. This prilgrimage was also linked to the December solstice
and to the passage of the sun through Zenith on October 30, when priests
visited the source of the Vilcanota river because the Sun was born
there, a pilgrimage referred to above.

Concerning the third theme, I will argue that the custom of
inviting each other was not done only in October, as the myth says, but
during the whole time from August to December when the primary concern of
people was in inducing fertility of land, llamas and humans. The great
feast when people would not quarrel and would invite each other was in
September, the Situa or Coya raymi, "the feast of the Queen, women and
the Moon". Then also, warriors representing the different ayllus of
Cuzco would drive out evil, each along his own ceque (Zuidema 1964; M.S.)
Guaman Poma (p. 251, 253) mentions that the same custom of inviting each
other was already done in August during early planting. The customs in
October, November and December can also be seen in this light. October,
the feast of Uma raymi -in Aymara, uma raymi means "the feast of Water
(uma)" -was distinguished (Guaman Poma p. 254, (256), 255 (257)), by
people and black llamas weeping for the coming rains. In November the
first preparations were made for the initiation rituals. Then the initiants would dance with old men, expressing their unity and identity. It
was also the month when the noble men from Cuzco would join the king in
the Itu ritual, expressing their common concern about the weather in the
next months. Besides its celebration in November it was carried out in
times of stress, of natural calamities due to "acts of God", when people
needed to unite most (Polo 1916 pp. 24-26). Although the initiation
rituals in December expressed a great individual competitiveness among
the initiants, it crossed the boundaries of family and ayllu and in that
sense united them all. Only in January, as the first act of the month
Capac raymi Camay quilla, after the December solstice, the boys of the
moieties Hanan Cuzco and Hurin Cuzco would fight each other in a ritual
battle. It was done with green, hard cactus fruits. The symbolic
meaning of the fight and of the fruits may have been that now no growth
or seed-setting was needed anymore, but what was needed was the process
of ripening, coinciding with the diminishing rains. If the ritual of
inviting each other in our myth does not say anything special about
October -only how it partakes in the ritual of all these months-, its

reference to the Thundergod may do this. Black llamas were starved in
October for this god to send his rains.

The next theme, no 4), makes even more clear that we are
dealing with data referring to the month of October. At three other places in the chronicles the same or similar data about women iñaca are mentioned. First Murua himself elsewhere (Murua 1962 book II ch. 13)
repeats his information about women called iñaca (erroneously written as
yucana) each having 100 or 150 servants. In this chapter they are called
the wives of governors of provinces, given to the latter by the king.
The specification of their rank is part of a detailed description of the
Incaic sumptuary laws on dress and adornments allowed to these governors.
We can see the myth then as a special reference to these people. Guaman
Poma (p. 740(754)) indicates that iñaca was the title of the wives of the
chiefs belonging to the "Incas-by-privilege" of Cuntisuyu (IV),
non-Incaic chiefs living around Cuzco in this direction who had been
given the rank and title of Inca. We recall that Cuntisuyu is the direction that includes the month of October in my proposed calendric use of
the ceque system (see figure 3). Guaman Poma confirms the kind of
descriptions of iñaca given by Murua. He narrows down the connection to
Cuntisuyu (IV). Finally, Sarmiento de Gamboa in 1572 mentions how
Pachacuti Inca, the ninth king; (in the myth of Murua we met him as the
father of Cusi Huanachiri):

> "had four legitimate sons in his wife Mama Anaguarqui, (and
> that) he had one hundred bastard sons and fifty daughters
> whom, because they were so many, he called Hatun aillo, which
> means "the large lineage". This lineage is called by another
> name Inaca panaca aillo".

Mama Anahuarque was a woman who came from the villages Choco and Cachona,
in the valley just South of Cuzco. The mountain called Anahuarque
belonged to these villages. Mama Anahuarque as a mythical ancestress was
worshipped in the valley by the non-Incaic inhabitants and especially by
the people of the village of Uma, the presentday San Jerónimo, who
celebrated their initiation rituals in the month of October, Uma raymi.
Mountain Anahuarque played an important role in the initiation rituals of
the noble youths in December and for the people of Uma in October. The
Inca boys would run in competition down the mountain with a female llama,
while their girl-helpers had gone already before in order to receive them
with cornbeer. The race was done in memory of the Flood when only

mountain Anahuarque had survived by rising above the waters. On the
basis of these and other data the name of Anahuarque can probably be
explained from Aymara —a language at some time with influence in the
valley and especially in the village of Uma— referring to "bringing boys
and girls together for marriage" (Zuidema & Urton 1977). It is remarkable how Murua and Sarmiento can give different information about who was
the founder of Iñaca panaca, but that both keep to the numbers 100 and 50
or 100 and 150. Moreover, Sarmiento, by referring to Mama Anahuarque,
supports the calendrical and astonomical context. The Southern Cross and
α and β Centaurus rise in Cuntisuyu (IV) behind mountain Anahuarque, in
the direction of ceque Anahuarque, IV_B 3 b. The mythological and ritual
context is similar to that of the celestial Black Llama in Huarochiri.
The period of lower culmination of α and β Centaurus is, as we saw, from
about October 7 to November 2, when in Hurochiri the celestial Black
Llama prevented the Flood by drinking the waters of the Earth, and when
in Cuzco real black llamas were starved in order to weep for rain. Thus,
Murua's myth, correlating the theme of the women iñaca to the month of
October, shows a remarkable consistency with the calendrical use.

In terms of the argument developed here, the last theme of
Murua's myth, Cusi Huanachiri's equation to Manco Inca, now also becomes
of major interest for the calendar. Other chroniclers, like Sarmiento
(ch. 16) mention as brother of the third king Lloque Yupanqui, not Manco
Inca, but Manco Sapaca. His name in Aymara means "ancestor semen". A
myth (Zuidema 1964 pp. 133-139, 1981b; Duviols 1979) about him mentions
how he sought a wife for Lloque Yupanqui in the village of Uma. Lloque
Yupanqui had to procreate a son when, because of his old age, he was not
thought able to do this anymore. His son, king Mayta Capac, was born
with teeth, grew up in a year and when he was two years old he defeated
boys of the non-Incaic ayllus in Cuzco with his sling, thereby subduing
the whole original population to the Incas. Mayta Capac was like the
Thundergod in the myth of Central Peru who, after being nurtured at
sunrise by the partridge-woman, immediately grew up and defeated his enemies with his sling. Other data confirm the association of Mayta Capac
and of Pachacuti Inca to the Thundergod.

If now I am permitted to give a composite picture of my
interpretation of these myths in terms of the calendar, we could argue as
follows. Lloque Yupanqui, whose panaca or group of descendants belongs
to II 3, is related to the weak sun of the June solstice; that is to the

death of the old sun and to the birth of the new sun. He is forced to marry a woman from Uma, the village that celebrates its initiation rituals in October. The man who forces him, his brother Manco Sapaca or Cusi Huanachiri, represents the new fertility in man at a time when also the fertility of the Earth renovates. He, or Pachacuti Inca -or Mayta Capac probably also -attract by their hundred sons alliances to the surrounding non-Incaic people. They celebrate   the initiation rituals in Cuzco after the same rituals of Uma in October and after October 30 when Cusi Huanachiri drinks to the victorious sun who then goes through Zenith. Cusi Huanachiri invents the ritual of ear piercing; in fact he adopts it from the people who live towards where the Sun is born. This ritual is the last one of the initiation rituals, carried out during the December solstice. It is done in the plowed fields and near sources of water, reconfirming the ties of the initiated boys to the land and to the established political order.

Even if the anthropological argument given here had to be condensed extremely, it demonstrates the importance of Murua's myth for our calendrical reconstruction. It shows how the data for the sidereal lunar calendar have to be found in terms of political organization and dynastic mythology. Murua relates the myth of Pachacuti Inca, preceding the one discussed here, to the months of Capac raymi and Capac raymi Camay quilla. An equally strong argument can be made that this myth is also calendrically correct (Zuidema M.S.c.) Eventually perhaps we will be able to account for the whole calendar in terms of these dynastic and political myths.

AN INTEGRATED VIEW OF THE DOUBLE FUNCTION OF THE CEQUE SYSTEM

My analysis of the myth of Murua, taken in the context of the other myths, leads us back to the initial problem that allowed us to discover an essential characteristic of the ceque system: its dichotomy of being, a) on the one hand, a system of coordinates -giving a geographic description and one of astronomical rising and setting points-, and b), on the other hand, a giant quipu-calendar reproducing real quipus used for calendrical purposes. Now we can reintegrate these two functions again, showing how in the case of two ceques these referred to the same time of the year, in terms of both astronomical coordinates and in terms of the quipu-calendar. Let me give first various data to be used for this purpose.

Ceques IV_B 3 b and IV_A 3 a,c. The interest in the quasi-circumpolar
constellations allowed the Incas to integrate the two properties of the
ceque system and to give them a single base for technical development.
In terms of directions, the direction of ceque Anahuarque, IV_B 3 b -the
first ceque of IV, counting for the purpose of the argument here the
ceques of IV from this one counter clockwise- can be defined precisely
towards 34° East of South as various huacas on this and neighbouring
ceques can be located, some having preserved their old names. Within
this direction lies the area of rising for the Southern Cross and α and β
Centaurus, their joint axis of rising being some 30° East of South.

The area of their setting lies in IV_A, just within the limits
of ceque IV_A 3 a,c. In this case the proof for this statement is
somewhat more complicated but not less conclusive. The Incas desired to
make the number 41 (which was of astronomical interest to them in the
sidereal lunar calendar) an integral divisible number, so they composed
one ceque onto two original ceques, making the total 42 (2 x 3 x 7). On
the ground these two ceques were separate. Their directions can be
defined sharply. IV_A 3 c begins outside Cuzco but within the valley and
has as its last huaca, the one that defines its direction, the seventh of
IV_A 3 a,c, Cachicalla, which "is a gorge in between two mountains like a
gate". This mountainpass is still known under that name and is on the
modern road to the South of Cuzco. The direction of Cachicalla is
4° West of South. Ceque IV_A 3 a begins in the neighbourhood Cayaocachi
of Cuzco itself with the eighth huaca, Quiacas amaro of the joint ceque.
Here I could locate quite exactly the 11th huaca of IV_A 3 a,c. It is the
mountain Cumpi, visible from the temple of the Sun, Coricancha, in the
direction of 41° West of South. Its distance from Coricancha is about
where IV_A 3 c starts. Ceque IV_B 3 b and IV_A 3 a,c are the ceques with
the highest number of huacas, 15, the next ones being II 1 a (13 huacas),
I 1 b (13 huacas) and I 1 a (12 huacas). The ceque named Anahuarque is
moreover the only ceque that starts with a huaca within the temple of the
Sun; huaca Sabaraura, in the ceque system said to be located "where is
now the window ("mirador" in Spanish) of Santo Domingo". This is just
behind the famous curved Inca wall of Coricancha and a place where the
Incas also had a window in the wall. When one early chronicler
(Santillan 1950 p. 47), discussing the division of the four suyus
extending from Cuzco, says that the town itself belonged to Cuntisuyu
(IV), the other suyus starting beyond it, he probably over-interprets the

fact that Coricancha, the center of town itself, was tied to ceque Anahuarque.

Two other data may be of importance to round off this general picture of Cuntisuyu, and to make our argument for integrating the two functions of the ceque system. Besides ceque Anahuarque, the only other ceque that refers to data on an important female hero in Cuzco is ceque IV_A 3 a,c. The first huaca, called <u>Tanancuri cota</u> "was a stone into which people said a woman had been converted who came with the <u>pururaucas</u>". Tanancuricota, also called <u>Chanancuricoca</u>, was a woman who helped Pachacuti Inca to defend the city against the South when it was attacked by the people of the Chancas. She then turned into stone (Sarmiento ch. 27; Santacruz Pachacuti P. 238).

<u>Sidereal months of 28 nights</u>. The other data refer to a somewhat different way of counting the huacas in IV, revealing a greater importance in terms of the sidereal lunar counting. In August, corresponding to II 1, the passage of the setting sun through the four pillars on mountain Picchu served to establish its relationship to the moon. Even today the first 15 days of August are used in Cuzco, not only to observe the phases of the moon during this time (especially at the beginning and end of it), but also to observe its position in the sky and its relationship to the Pleiades (Lira 1946 p. 18-19; Zuidema 1982a). If we can relate the counting of ceque II 1 a and its 13 huacas to the second half of August, and thus probably to the time that the sun was in-between the central and the southern pillars, then we can define two periods of 28 nights, representing sidereal months, from the time that the sun passed the central pillars by joining II 1 a + IV_B 3 b, and IV_B 2 + IV_B 1. We can suggest that the first of these sidereal lunar months was used for defining in general a period of 28 nights, observing how the moon would recur close to the same position it had before. The second of these months then would correlate it more sharply to the movement of the Southern Cross plus α and β Centaurus by way of the periods of 4, 4, 5, 5, 5, 5 nights, the number of huacas given in sequence on the ceques by Cobo. The period of IV_A 3 a,c with its 15 huacas plus IV_A 2 and IV_A 1 with their 22 huacas, would help to correlate the moon to the Pleiades, using especially the latter periods of 3, 4, 4, 3, 4, 4 nights for that purpose. Even if this scenario of events cannot be proven completely with empirically-observed data it makes us critically aware of the necessity of asking the question how the sidereal lunar connection was observed; of

the importance that one name of the celestial Llama, catachilla
unuchilla, refers to the concept of movement of the sky; and to the fact
that ethnographic data probably still can be found to solve the problem.

The double function of ceques IV$_B$ 3 b and IV$_A$ 3 a,c

Ceques IV$_B$ 3 b and IV$_A$ 3 a,c helped to define in terms of
space, by way of indicating the rising and setting points of the Southern
Cross and α and β Centaurus, the period from their first heliacal set on
September 7 to their last heliacal rise on November 2. They indicated
the same period by their use in the quipu-calendar; from the first use of
IV$_B$ 3 b on September 3 to the last use of IV$_A$ 3 a,c on November 2, with
the use during the time in between of the ceques of IV$_B$ 2 and IV$_B$ 1. It
was the further use of the moon in the months of Coya raymi, dedicated to
the moon, the queen and women in general, and of Uma raymi, dedicated to
the first rains, that allowed to define this period sharply in fixing it
into the solar calendar.

Peoples in Micronesia used the rising and setting points of
the Southern Cross plus α and β Centaurus for defining the South
(Goodenough 1953) and one chronicler says the same for Peru (Calancha
1639 vol I. p.50). Cuzco anchored its ceque system to the two ceques
that defined the South. They defined also their concept of origin. Not
only were these ceques related to female heroes; but also the cave from
where the first Incas were born, Pacarictambo "the cave of night, birth
and emergence", was in that direction. We noticed the concepts of inver-
sion related to IV in a symbolic and in an astronomical sense. In IV the
two uses of the ceque system were combined and from here they went their
different ways. It is only here that one and the same ceque can indicate
the same property of the ceque system in a spatial and in a calendrical
sense. This may have been the reason why a myth like Murua's could exist
and why it could give us the confirmation and clue of how to read the
ceque system as a quipu-calendar. It may also be the reason why the numbers of ceques II 1 a through IV$_B$ 1 might give us the clue to how
sidereal lunar months of 27 1/3 or 28 nights actually were observed.

CONCLUSIONS

The reconstruction of the sidereal lunar calendar of the Incas
in Cuzco was made possible 1) because the ceque system as a system of
directions gives us precise information about key astronomical observations; 2) because the ceque system was also used as a quipu-calendar;

and 3) because information about Inca rituals and myths confirms the place of the sidereal lunar count of 328 nights from June 9 to May 3.

Observation of the four pillars on mountain Picchu enabled the Incas to define the day when the sun sets in the anti-Zenith position and a monthlong period around that date of August 18. With this observation they fixed within the solar year a synodic lunar year of twelve months counted from June 6, starting with three days of invisible moon. The actual synodic lunar months in a given year were measured against the fixed synodic lunar calendar. The first possible visible new moon of the month during which the June solstice can be observed occurs as early as May 26. Coricancha, the temple of the sun, is aligned to sunrise on this day. With a full moon on the June solstice, a new moon coincides with the first heliacal rise of the Pleiades around June 5-8 (in 1500 A.D.). The first sidereal lunar half year ended on November 21, just after the first heliacal set of the Pleiades; the second sidereal lunar half year ended some two weeks after the last heliacal set of the Pleiades. The observations made from Coricancha toward Pleiades rise and set give evidence that the Incas intended to see the rise in relation to sunrise on May 26 and the set in relation to sunset on April 26, and thus that they intended to relate the sidereal lunar year in a precise way to the solar year and to the synodic lunar year.

With this understanding of the place of the sidereal lunar calendar in the solar year, we receive a strong support for the ethnohistorical data on Incaic calendrical concerns in their rituals. December solstice and June solstice were the most important celebrations in Cuzco carried out by the state. The one month of the small feast of the sun, included the June solstice; the December solstice, the great feast of the sun, was, however, celebrated in two months, one before and one after. The quipu-calendar explains how these interests were given calendrical form.

Even if certain feasts were in honor of the sun, they were celebrated during full moon. This is especially true of the (Incaic) months of August, September, January, April and May. The interpretation of the quipu-calendar supports the ethnohistorical data.

The most important celebrations of the stars were of the Pleiades -from the time when they disappear to the time that they reappear and during the time of their upper culmination around midnight- and of the Southern Cross with α and β Centaurus, when these constellations have their lower culminations around midnight. The counting of these last lower culminations followed by the upper culmination of the Pleiades was accomplished in the quipu-calendar by Cuntisuyu (IV). Ceques IV_B 3 b and IV_A 3 a,c in this suyu (IV) defined the same period by indicating resp. the rising and setting points of the Southern Cross with α and β Centaurus. The myth of Cusi Huanachiri confirms the double function of the ceque system here; an interpretation of its use of the date of October 1 proves this to be correct.

The Incas used the two groups of constellations —the Pleiades on the one hand and the Southern Cross with α and β Centaurus on the other, seen as being opposed to each other in the sky— for constructing a sidereal lunar calendar of 328 nights. The example of an Indonesian calendar —of Acheh in Sumatra (Snouck Hurgronje 1893 Vol. 1)- demonstrates a similar procedure. But while here each of the twelve synodic months, that were counted from a given date in the solar year, was paired to one of twelve sidereal months of 27 or 28 days, the Incas used a different system. They accommodated in their count of the twelve periods of the sidereal lunar year for other astronomical observations of the sun and the moon. Thus they arrived at a system of twelve irregular periods within their sidereal lunar year of 328 nights. Further study may demonstrate this to be an original contribution of Andean civilization to our knowledge of calendars in general.

ACKNOWLEDGEMENTS

I thank the following organizations for their support over the years of this research: National Science Foundation, Social Science Research Council, American Council of Learned Societies, University of Illinois and the Organization Earthwatch that, with the enthusiastic collaboration of its supporters in the fieldwork, helped the work of Aveni and me in the years 1976, 1977, 1979 and 1980.

REFERENCES

Anonymous   (1879(1539)). Relación del Sitio del Cuzco y Principio de
               las Guerras Civiles del Perú hasta la Muerte de Diego
               Almagro. In Colección de Libros Españoles Raros y Curiosos,
               25 Vols. Madrid: 1871-96.
Anonymous   (1908(?)). Discurso de la sucesión y gobierno de los Yngas.
               In V. Maúrtua, Juicio de Limites entre el Perú y Bolivia,
               Vol. 8, Chunchos. Lima.
Aveni, A.F.   (1981). Horizon Astronomy in Incaic Cuzco. In Archaeoas-
               tronomy in the Americas, Ed. R.A. Williamson. Los Altos,
               Cal.: Ballena Press.
Avila, Francisco de. See: Taylor, G.
Bertonio, L.   (1956(1612)). Vocabulario de la Lengua Aymara. La Paz: Don
               Bosco.
Betanzos, Juan de (1968). Suma y Narración de los Incas (1551).
               Biblioteca de Autores Espanoles. Crónicas de Interés
               Indigena, Madrid.
Calancha, A. de la (1639). Coronica Moralizada del Orden de San Agustín
               en el Perú. Barcelona.
Cieza de León, Pedro de (1945(1551)). La Crónica del Perú. Buenos
               Aires: Espasa-Calpe Argentina, S.A.
Cobo, Bernabé (1956(1653)). Historia del Nuevo Mundo. Madrid:
               Biblioteca de Autores Españoles.
Duviols, P. (1974-76). Une petite Chronique retrouvée: Errores, Ritos,
               Supersticiones y Ceremonias de los Yndios de la Provincia de
               Chinchaycocha y Otras del Pirú. Journal de la Société des
               Américanistes 63, 275-297.
Duviols, P. (1978). Camaquen Upani: un Concept Animiste des Anciens
               Peruviens. In Amerikanistische Studien I, eds R. Hartmann &
               U. Oberem. pp. 132-144. Collectanea Instituti, Anthropos,
               Vol. 20. D-5205 St. Augustin 1: Haus Völker und Kulturen,
               Anthropos Institut.
Duviols, P. (1979). La Guerra entre el Cuzco y los Chanca. ?Historia o
               Mito? In Economía y Sociedad en los Andes y Mesoamérica, ed.
               J. A. Franch, pp. 363-371. Revista de la Universidad Complu-
               tense XXVIII 117. Madrid: Editorial de la Universidad Complu-
               tense.
Goodenough, W.H. (1953). Native Astronomy in the Central Carolines.
               University Museum. Philadelphia: University of Pennsylvania.
Guaman Poma de Ayala (Waman Puma), Felipe (1980(1583-1615)).
               El Primer Nueva Corónica y Buen Gobierno. Eds. J.V. Murra &
               R. Adorno. México D. F.: Siglo XXI.
Herrera, P. (1916). Apunte Cronológico de las Obras y Trabajos del
               Cabildo y Municipalidad de Quito desde 1534 hasta 1714
               (Primera Epoca), Vol. 1. Quito.
Lira, J. A. (1946). Farmacopéa Tradicional Indígena y Prácticas Rituales.
               Lima.
Maass, A. (1924). Sternkunde und Sterndeuterei im Malaiischen Archipel.
               Tijdschrift voor Indische Taal-, Land en Volkenkunde, 64,406-408.
Meza Bueno, J.B. (1943). La Fiesta de San Juan. Vol. 2, no.20 of the
               Monografías de Geografía Humana, catalogued by J. Cornejo
               Bouroncle. Cuzco, Perú.
Molina, C. de (1943(1573)). Fábulas y Ritos de los Incas. Lima: D.
               Miranda
Murúa, Fray Martín de (1946(1590)). Historia del Origen y Genealogía
               Real de los Reyes Incas del Perú. Ed. C. Bayle S.J.
               Bibliotheca "Missionalia Hispánica". Madrid: Instituto Santo
               Toribio de Mogrovejo.

Murúa, Fray Martin de   (1962(1611-1618)). Historia General del Perú,
    Orígen y Descendencia de los Incas...Prol. Duque de Welling-
    ton, Introduction M. Ballesteros-Gaibrois. Madrid: Coleccion
    Joyas Bibliográficas, Bibliotheca Americana Vetus I.
Pizarro, P. (1978(1572)). Relación del Descubrimiento y Conquista de los
    Reinos del Perú. Ed. G. Lohmann Villena, Note by P. Duviols.
    Lima: Pointifícia Universidad Católica del Perú.
Polo de Ondegardo (1916(1584)). Los Errores y Supersticiones de los
    Indios Sacados del Tratado y Averiguación que hizo el
    Licenciado Polo. Eds. H. Urteaga & C. Romero. Lima.
Rowe, J.H. (1979(1653)). An Account of the Shrines of Ancient Cuzco.
    Nawpa Pacha 17, 1-80. Berkeley: Institute of Andean Studies.
Santacruz Pachacuti Yamqui Salcamaygua, Joan de (1950(1613)). Relación
    de Antigüedades de este Reyno del Perú. Reproduction of
    edition M. Jimenez de la Espada. Tres Relaciones de Anti-
    güedades Peruanas, pp. 207-281. Asunción de Paraguay:
    Guaranía.
Santillan, F. de (1950(1563)). Relación del Origen, Descendencia,
    Política y Gobierno de los Incas. Reproduction of the edition
    by M. Jimenezde la Espada, Tres Relaciones de Antigüedades
    Peruanas, p. 35-131. Asunción de Paraguay: Guaranía.
Snouck Hurgronje, C. (1893). De Atjehers. Batavia (Djakarta).
Taylor, G. (1974-76). Camay, Camac et Camasca dans le Manuscrit Quechua
    de Huarochiri. Journal de la Société des Américanistes,
    LXIII, 230-244.
Taylor, G. (1980(1608)). Rites et Traditions de Huarochiri. Paris:
    L'Harmattan.
Urton, G. (1981). At the Crossroads of the Earth and the Sky. An
    Andean Cosmology. Austin: University of Texas Press.
Zuidema, R.T. (1964). The Ceque System of Cuzco. The Social
    Organization of the Capital of the Inca. Leiden: E.J.
    Brill.
Zuidema, R.T. (1973). La Parenté et le Culte des Ancêtres dans Trois
    Communautés Péruviennes: Un Compte-Rendu de 1622 par
    Hernandez Príncipe. Signes et Languages des Amériques,
    Recherches Amérindiennes au Quebec, Vol. III, nos. 1-2, pp.
    129-145. Montréal.
    also published as:
    Kinship and Ancestorcult in Three Peruvian Communities;
    Hernandez Principe's Account in 1622. Bulletin Institut
    Francais des Etudes Andines, Vol. II, pp. 16-33. Lima.
Zuidema, R.T. (1977). The Inca Calendar. In Native American Astronomy,
    Ed. A.F. Aveni, pp. 219-259. Austin: University of Texas
    Press.
Zuidema, R.T. (1979). El Ushnu. In Economía y Sociedad en los Andes y
    Mesoamérica, ed. J.A. Franch, pp. 317-362. Revista de la
    Universidad Complutense XXVIII 117. Madrid: Editorial de la
    Universidad Complutense.
Zuidema, R.T. (1981a). The Inca Observations of the Solar and Lunar
    Passages through Zenith and Anti-Zenith at Cuzco.
    In Archaeoastronomy in the Americas, Ed. R.A. Williamson.
    Los Altos, Cal.: Ballena Press.
Zuidema, R.T. (1981b). Myth and History in Ancient Peru. In The Logic
    of Culture, Ed. I. Rossi. South Hadley, Mass.: J.F. Bergin
    Publ., Inc.

Zuidema, R.T. (1982a). CATACHILLAY. The Pleiades and the Calendar of the Incas. Conference on Archaeoastronomy in Native America, New York Academy of Sciences, April 1981. To be published in Acts of the Conference by the Academy.

Zuidema, R.T. (1982b). Las Pleyades y la Organización Política Andina. Paper given at "Jornadas de Etnohistoria", Lima, Peru, May 1981. To be published in Acts of the Asociación Peruana de Ethnohistoria.

Zuidema, R.T. (M.S.a.). Bureaucracy and Systematic Knowledge in Andean Civilization. In Symposium on "Native American States and Indianist Policy; Historical Consciousness of the Incas and Aztecas, 1400-1800.", Stanford University, December 1978. To be published in Volume of the Symposium.

Zuidema, R.T. (M.S.b). The Myths of Inca Urco and Cusi Huanachiri.

Zuidema, R.T. (M.S.c). Puma and Jaguar in the Cosmology of Cuzco. To be published in Volume on Animal Symbolism.

Zuidema, R.T. & G. Urton. (1976). La Constelación de la Llama en los Andes Peruanos. Allpanchis Phuturinqa, 9, 59-119. Cusco, Peru.

ASTRONOMICAL NORMS IN MESOAMERICAN RITUAL AND TIME-RECKONING

G. Brotherston
Department of Literature, University of Essex

TIME-RECKONING IN HIEROGLYPHIC AND IN ICONOGRAPHIC TEXTS

Part of a larger study of Mesoamerican chronology (Brotherston,
1981, 1981 a, 1981 b), this paper focuses on native texts written in the
hieroglyphic and the closely related Mixtec-Aztec or iconographic script
typical of that region. Structurally these two script traditions share a
ritual system based on the 18 weeks or 'Fasts' of the year, each of 20
days, and on the tonalamatl or Sacred Round of 9 Figures (1' to 9'), 13
Numbers (also 'Fliers' or Quecholli), and 20 Signs (I-XX; Tables 1-2). In
hieroglyphic these sets are known calendrically to relate to the metric
year of 365 days; in iconographic script, however, while the metric year
is observed through the Day Count common to all Mesoamerica (tonalpoualli),
the calendrical or tribute year in question, tied as it is to agricultural
produce rather than labour, has to be identified as solar and is subject
to an annual day count inclusive of unnamed leap-days (cemilhuitlapoualli).
The conventions shared by these two script traditions are further shown to
include markers for larger time periods (Table 3); in particular, the two
principal periods observed in the year-calendar, the 52-year Round and the
400-year 'Head' (tzontli), find their lowest common multiple in the Era of
5200 years (see Rios p.8), which arithmetically matches the 5200 'years'
of 360 days found in hieroglyphic. In both cases the Era begins in 3113 BC
(including the year 0). In the iconographic tradition this corresponds to
the year 13 Reed specified on the Aztec Sunstone and the Cuauntitlan
Annals; in hieroglyphic it corresponds to the katun-ending 4 Ahau. Proof
of these claims is set out in A Key to the Mesoamerican Reckoning of Time
(Brotherston, 1981 b).

Intricately related to this overall chronology of the Era,
Mesoamerican astronomical data in iconographic and hieroglyphic script
alike show a primary concern with the readily visible bodies that move
along the ecliptic or zodiac road, which in the tropics passes through

zenith and nadir (Brotherston, 1976, 1976 a). These are sun, moon, and
the five planets, Mercury and Venus who 'plunge' through inferior conjunction or 'the underworld', and the superior planets Mars, Jupiter and
the slow old man Saturn who do not. About how sun and moon are represented there is no doubt, either among the glyphs of the hieroglyphic 'sky
band' or among the celestial figures who parade through the iconographic
texts. Less certain is how the planets may correspond to the remaining
glyphs in the sky band - modified Series III Signs that include Venus as
VIII (lamat) - or to the remaining celestial walkers (Fig. 1). At any
event the important fact here is that computations attaching to these
bodies, directly and indirectly, are numerical in the sense of non-
geometrical, and take a whole day as a minimum unit. As such they compare
readily with the cuneiform data from early Babylon as Neugebauer now presents them (1976:3): 'It is a historical insight of great significance
that the earliest existing mathematical astronomy was governed by numerical
techniques, not by geometrical considerations, and, on the other hand,
that the development of geometrical explanations is by no means such a
'natural' step as it might seem to us who grew up in the tradition
founded by the Greek astronomers of the Hellenistic and Roman period.'

In practice the deciphering of such numerical data in
Mesoamerica has so far been almost entirely restricted to the hieroglyphic
texts, where they are found in one of two main forms. On the one hand

Fig. 1: Possible planetary signs appearing with sun and moon
(Inferior planets). (a) hieroglyphic sky-band;
Palenque Temple of Cross panel, left half; Madrid,
p.12, right half; Paris, pp.23-4; Dresden, pp.51-8.
(b) Celestial walkers: **Borgia, p.55 (line 2)**; **Fejérváry**,
pp.30-32, 35-40 (cf. Nowotny, 1961:208)

	Mercury	Venus
(a)	Sign III akab ('Night'), wotan	Sign VIII lamat, kanil (Venus)
(b)	jaguar-skin cargo (= Night, Copan D B4)	plunging Quetzal bird of Venus ('12')

the screenfold books contain tables of bars (5) and dots (1), laid out
with vigesimal place value, whose unit groupings incontestably refer to
heavenly bodies; such are the heliacal Venus and the eclipse tables in the
Dresden Codex. The other type of data are those derived from calendar
dates within the Era, which point to astronomically significant moments.
Of this kind are the 'determinants' of the solar year decoded by Teeple,
and certain lunar-solar and planetary conjunctions patent enough to con-
vince the sceptical Sir Eric Thompson (1971:317, 196, 227, etc.). Upon
examination both types of data can likewise be discovered in the iconographic texts, where they mutually reinforce each other and their hieroglyphic counterparts.

      Formal equivalents to the hieroglyphic tables of bar-and-dot
numbers can be readily found in iconographic texts in the ritual genre,
specifically the Fejérváry and Cospi screenfolds. That these might have
an astronomical significance was early suspected by Seler (1901) who,
however, failed to clinch his case. More recently, in his indispensable
catalogue of the iconographic ritual texts, Tlacuilolli (1961), Nowotny
played down the likelihood of such a reading, reacting strongly against
the wholesale astrology of previous scholars. Yet that he overreacted is
suggested by the way he altogether underestimates astronomy as normative
in ritual. The 18 Fasts of the solar year go unacknowledged even in his
master text Borgia (pp.29-46, Lesser Tecuilhuitl to Etzalcualiztli); and
while he lists ritual chapters or 'almanachs' based on the four-year or
leap-day span, he does not identify analogous chapters based on the double
of this span, the 8-year octaeteris, doubled in turn to the 16-year span
over which deliveries of cloth tribute, due each equinox, total the teeth
of the upper and lower jaw (cf. Customs ff. 2v-6r). More seriously,
Nowotny doesn't even mention the set of eleven armed figures which
Customs of New Spain (ff. 49-59) connects with latitude and the 'drunken'
curve of the ecliptic and which relates numerically to the days, and to
the years of the lunar-solar epact. This 'zodiacal' eleven accompanies
the bar-and-dot numerals (with their unit maximum of 11) in Fejérváry and
Cospi and reappears in Laud (pp.38-44), while with reference to hieroglyphic sources Kelley has raised the possibility of an eleven-figure
zodiac at Chichen(Kelley1976:47-50; also, cf. Paris, pp.23-4 with the 11
divisions in Madrid,pp.13-18).Of the Thirteen Heroes it is the Night Lord
who is 11, and in Mendoza (pp.61-3) the calmecac is adorned by 11 stars,
which according to Clerke (1911) was once the total of zodiacal constellations in the Old World.

The particular reasons why tabular data in the iconographic
ritual texts have been less satisfactorily related to astronomy than those
in hieroglyphic will become clear below. They include a greater reliance
on the principle of shifting by means of ciphers between dimensions of
time (days and years, nights and moons); their greater attention to the
Pythagorean properties of numbers, their squares and cubes, their serial
totals (e.g. 4 as 1 2 3 4) and the principle of angle-value notation; and
the fact that they do not always appear as standardised and objectified
totals, distinct from writing on the one hand and picture on the other, as
in hieroglyphic texts, but may have to be extricated from a complex page
or figure design. The Fejérváry bar-and-dot pages which Seler believed
refer to Venus are said by Nowotny simply to represent prayersticks on
altars (Opfertisch) used for hunting and other propitiation ceremonies,
and are lumped under the all-purpose rubric 'Rituale mit Bündeln
abgezählter Gegenstände' (1961:272-4). In fact they are neither one nor
the other but both, according to the principle of multiple reading that is
no less than essential to ritual texts in iconographic script.

As for the second type of data, calendrical dates within the
Era, iconographic equivalents to the hieroglyphic examples could not be
recognised hitherto simply because the Era itself has generally been
thought exclusive to the latter tradition. Now that iconographic texts,
too, can be shown to respect the same base date of 3113 BC, analogous
datings may be found in them. A key text in this respect is the obverse
of the Vienna Codex, here called the Tepexic Annals, which counts out the
whole span from the Era base to year 2 House 1361 AD (p.50, the probable
date of composition), and projects to a close exactly 4800 years or 12
Heads from its first named year (5 Flint 3108 BC to 8 Flint 1692 AD; the
correlation used by this and geographically related texts from Anahuac, by
which 3113 BC equals 13 Reed and which makes 1516 AD equal 1 Flint, is
here referred to as 'Cl'). Among the Tilantongo or 'C10' group of texts,
the reverse of the Zouche Codex, a biography of Eight Deer, confirms the
fire-kindling performed by that hero in 1047 AD as a 'Jade Round' or 80 x
52 years from 3113 BC (p.35); among the Tenochtitlan and C11 texts the
Mexicanus (p.9) specifies 1570 AD both AD and as 3 x 29 x 52 plus 158
years from 1 Flint 3112 BC, 4682 years in all (cf. Table 6). Also, within
this Era framework with its three main correlations of Series III years
(Cl, C10, C11), ritual year multiples are frequently used, in the counting
out of dates, which themselves can be shown to be astronomical in origin.

LUNAR-SOLAR AND SYNODIC PERIODS

The bar-and-dot sequence proper to the eleven figures in Fejérváry is laid out as follows (pp.5-14):

11	6	8	5	11	11	10	7 8 9	8	11	11	7	7	
x 31	9	24	30	30	1	30	9 9 9	27	33	33	27	27	(1)
341		396		330	11	300	216	216	363	363	189	189	

The total is a quite unambiguous 2914. This approximates the well-established 2920 which in Dresden (pp.46-50) serves as the lowest common multiple of metric solar and Venus years, i.e.: 8 x 365 = 5 x 584 = 2920; and this encouraged Seler to relate it to those bodies, though he could not account for the missing 6. Just such a connection may be made, however, and with the moon as well, when the 2914 total is related to the true lunar-solar octaeteris (as opposed to metric solar and Venus years) and is read as follows:

$$2914 = \text{8 solar years less 8 days at 365.24 d. per year}$$
$$= \text{5 Venus years less 5 days at 583.92 d. per year} \quad (2)$$
$$= \text{99 moons less 9 days at 29.53 d. per moon}$$

This reading accords well with the known ritual priorities of the year calendar. For, commemorated in the Atamale ceremony, the octaeteris falls into 'leap-day' halves as 4 + 4 years, which as a cipher echoes the 4 + 4 days it officially takes Venus to enter and leave the underworld at inferior conjunction, between west and east. And just this correspondence between 4 + 4 years and days lies at the heart of funeral liturgy practised throughout North America, whereby the days of the mourners at the wake equal the 'years' of the soul on its way through the underworld (Brotherston, 1979:104). As for the moon, its affinity with the number 9 harks back through the Ennead or Night Lords to the tonala-matl itself, which exactly measures pregnancy from the first missed menses to birth as 9 moons of 29 nights (less one). Indeed, in this midwife's count, which also equals the body's orifices, each of the 9 Night Lords commemorates a monthly stage of gestation from the Fire of coition (1') to the Rain-god's full water-bag (9'). So far from being anomalous, the 'reduced' total of 2914 is a super-number of the kind analysed by Aveni (1981) and expresses the octaeteris through ciphers thoroughly characteristic of sun, moon and Venus and of the ritual attaching to them.

In Fejérváry, the 2914 total formally completes the first of two sequences of bar-and-dot numbers. The transition to the second is marked visually by a new page format while the optimum number base doubles from 11 to 22. As for the accompanying characters, the team of eleven is succeeded by the 18 Fasts of the solar year beginning with lesser Micailhuitl; so that in the process, over the full 18 pages (pp.5-22) and beginning with Tititl, the eleven double as the Fasts prior to this start. The Fast dedicated to second burial, as the equivalent depiction in Laud (p.21) makes clear, Micailhuitl specifies the creatures that accompany the soul after death, according to a belief still current among the Navajo: Turkey and Dog, the domestic members of the thirteen Quecholli ('9') and the Twenty Signs (X) respectively (cf. Brotherston, 1979:112). Heading the second sequence of numbers in Fejérváry, the Micailhuitl pages also introduce new design conventions relating to the dot, which now may be rendered hollow or dispensable, and to the bar. For the sheer layout of the fourth subset of numbers (p.16) magnifies the bar 5 to its square and the higher powers (5, 5^2, 5^3, 5^4). And all four subsets on pp.15-16 involve serial totals based on 5 (5-9, 11-14; 5-14, 18-21) as well as illustrating the principle of $90°$ angle value by which 5 becomes 50, as in the hand-signs of our antiquity, equal to Γ in Greek and L in Latin. Both these operations are exemplified in the two opening symbols of the Fast: the first confirms Eagle as its own serial or 'sigma' total, this bird being both '5' of the Quecholli and XV of the Twenty Signs ('sigma' 5 or 1 2 3 4 5 = 15); the second turns the bar through $90°$ (cf. Fig. 2 and Table 5b). When used in hieroglyphic this angle-value converts days to uinals, vigesimally; here, in the iconographic tradition it is shown to be decimal by its attachment to the Sign for the domestic Dog (X) and by the fact that within the sigma totals on these pages the right-angle numbers

Fig. 2: Arithmetical indicators, Fejérváry, p.15

15 or 'sigma' 5 (Eagle feather,   5 plus decimal 5 (bar,
Sign XV, with three bars or   bar at $90°$)
5 x 3; note the 15 notches)

are integrated with ordinary numbers repeated ten times, to the same
effect.

Read in these terms, the pages devoted to the Fast
Micailhuitl at the start of the second number sequence in Fejérváry have
five subtotals laid out as follows:

$$p.15: \quad \sum_{r=11}^{14} (3 \times 10r) - 12 (3 \times 10) = 1140$$

$$\sum_{r=5}^{9} 10r = 350$$

$$p.16: \quad \sum_{r=18}^{21} 10r - (20 \times 10) = 5 \times 116 = 580 \tag{3}$$

$$\sum_{r=5}^{14} (r^2 - 4r) + 20 = 5^4 = 625$$

$$p.17: \quad (5 \times 20) + 9^2 = 181$$

Taken together, these five subtotals yield an overall 2876, the key to
which can be sought in the slightly higher total of the whole first
sequence, 2914. For

$$2876 = 1140 + 350 + 580 + 625 + 181$$

$$= 5 \times 5 \text{ Mercury years less 19 days} \left. \right\} \tag{4}$$
$$2914 = 5 \times 5 \text{ Mercury years plus 19 days} \quad \text{at 115.8 days per yr}$$

Though out by more than a day (0.077 per year), this five-part formula on
pages devoted to the number 5 has the advantage of bringing out the fifth-
of-a-fifth ratio that Mercury bears to the other inferior planet Venus
within the octaeteris; and that Mercury is its subject is intimated by the
third subtotal of 5 x 116 or five Mercury years. Other examples of this
116 total occur in the upper zodiac facade at Chichen as 4(13 + 16); in
Aubin ms 20 as dots on maguey paper (see below); and in Laud as dots of
fermented maguey, reading 19 36 31 13 10 and 7 (Sign VII; pp.31-38). As
Thompson notes (1971:215) re the 819 or 7 x 117 day cycle in hieroglyphic,
the Mercury year also approximates the product of the two primary ritual
sets 9 and 13.

The other advantage of this Mercury formula is that it highlights the number 19, which proves essential to this second sequence of
bar-and-dot numbers as a whole. For running from lesser Micailhuitl to
Atemoztli (where the angle-value is doubled, p.22), this sequence yields
the smaller and larger totals of 6940 and 7153, i.e. without and with the
hollow dots. The page subtotals read (pp.15-22):

1	1140	625	389	286	440	500	100	40
10	35	57		30	16		19	35
100								15
	1490	1195	389	586	600	500	290	1890	= 6940

hollow
dots: (5)

1			9	4			
10		1		6		2	1
100						1	213
		1205	398	650		310	2000	= 7153

These link sun, moon and Saturn as follows:

$$6940 = 19 \text{ solar years} = 235 \text{ moons}$$
$$7153 = 19 \text{ Saturn years less 19 days at } 377.5 \text{ days per year} \qquad (6)$$
$$= 6940 + 116 + 88 + 9$$

Just as the 2914 total of the first sequence of numbers concerned the moon
and the plungers, so these totals of the second sequence concern the
slowest planet; equally, just as the former gave the shorter lunar-solar
cycle of 8 years, the octaeteris, so the latter gives the longer cycle of
19 years, the Metonic. Hieroglyphically the Metonic cycle appears in the
7th century stelae 3 and A at Copan, in a formula read by Thompson as
follows (1971:195-6):

$$6940 = (19 \times 360) + (5 \times 20) = 20(360 - 13)$$
$$= 19 \text{ solar years} = 235 \text{ moons} \qquad (7)$$

Hitherto it had not been surely identified in iconographic texts.

Just as the octaeteris relates to Sign VIII (Venus or Rabbit),
so the 19-year cycle is designated calendrically by a special 'nosed' form

of Sign XIX (the Rain-god's mask); the hieroglyphic Sign XIX (cauac) likewise accompanies the Metonic formulae at Copan (7).   In the histories of
Tilantongo and the Mixteca, the damp 'Cloud-land', 19 years or a Rain
separates Eight Deer's birth (1011) from his father's first marriage (992),
which date stands 19 x 19 years or a Rain of Rains from the epochal Round
72 (631) and from the Tehuacan invasion and dynastic shift around 1353.
This last interval is shown in Rains at 1011 on the reverse of Vienna
(p.7) as 1 (elapsed) plus 6 x 3 (to come).   All this further confirms
Caso's readings (1967); and it highlights the favouring of given dates and
intervals in native history because they have resonance in the sky.
Counted out in leaves, 19 years is also the dominant unit in the 67-year
total given on the title-page map of the Mendoza Codex, which, opposed to
the 51 years around the margin of the same page, serves to record the 16-
year gap between C1 and C11 year names (13 Reed as 1359 or 1375).   The two
lots of 19 years appear in the southern and western quarters of this
tribute map; in the latter case it is shown to be 3 years more than the
16-year equinoctial-cloth span by the teeth in the toponym Tzompanco,
north, which in turn is 3 years more than the 13-year or quarter-Round
Reed (Sign XIII, east; Fig. 3).   The same three year multiples, 13, 16 and
19, appear together within the Fejérváry Metonic tables in the Fast
Quecholli (p.20).

The calendrical use of the 8-and 19-year lunar-solar cycles

Fig. 3:   Tribute quarters map, Mendoza p.1, showing year
totals of 67 (including the Metonic 19) and 51 (2 III to
13 XIII)

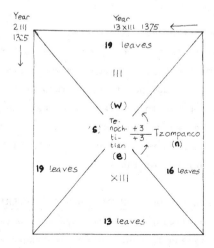

alerts us to the intermediary 11-year period likewise denoted by the
Twenty Signs as the Monkey, Sign XI; spelt out as such in the
Cuauhtinchan History (between 1340 and 1351 AD), the Monkey appears
together with the Rain in the centre of the Aztec Sunstone. Though not
quite so precise astronomically (136 moons being a day and half less than
11 years), the Monkey has the ritual advantage of matching in years what
the epact is in days (perhaps the logic underlying the '11 Monkey' design
at Xochicalco). In the same connection, the Twenty Signs also furnishes
the 10-year Dog, Sign X, to which the Monkey must eventually shrink due to
the lunar shortfall; Dog and Monkey are paired at the close of Vaticanus B
(pp.85-6) and the Popøl vuh (lines 8398-8410). Also, on the penultimate
page of Fejérváry (p.43), which may formally be seen as a continuation of
the Numbers chapter, the Monkey appears along with the Rain in an intricate bar-and-dot formula which shifts between dimensions of time, in this
case Fasts and years. In preparation for the final Series III date
embedded in the last icon page (p.44; cf. the equivalent bar-and-dot
sequence in Laud, p.44), this formula exemplifies how to move forward from
one year Series to the next, within a 260-unit tonalamatl of both days and
years:

$$
1\ VI \leftarrow 19\ x\ 2 \rightarrow \left\{ \begin{array}{l} 1\ I \leftarrow 11\ Fasts \longrightarrow 1\ II \\ 1\ I \leftarrow 11\ years \rightarrow \boxed{13\ I =} 1\ II \end{array} \right\}
$$

$$
= 1\ yr + 2\ Rains + 1\ yr + 1\ Monkey + 1\ yr = \quad 52 \tag{8}
$$

$$
\underline{208}\ (4\ Flames)
$$

$$
260\ (tonalamatl-\ yrs)
$$

For present purposes the year-name mechanism as such matters less than
function within it of the 19- and 11-year lunar-solar periods.

    In sum, the Fejérváry chapter comprising two extensive bar-
and-dot sequences (pp.5-22) points predominantly to an interest in lunar-
solar cycles, which prove to have a calendrical function in a variety of
texts, including the Fejérváry itself. And these cycles are specified
through formulae, of high ritual significance, which also serve to integrate the synodic periods of the inferior planets.

SIDEREAL PERIODS, OF THE MOON AND INFERIOR PLANETS

    We have already noted the parallels between the bar-and-dot
chapter in Fejérváry and the only other comparable chapter extant in

iconographic script, that in Cospi, pp.21-31. For they share the optimum
number base of 11 and the team of eleven figures, which however in the
case of Cospi govern the whole chapter, of 11 pages. Also Cospi gives continuous attention to emblems which attach to only certain of the 18 Fasts
in Fejérváry and which are known at least in part to refer to the zodiac.
For they appear in a zodiacal context in other iconographic texts as well
as in such Maya sources as Madrid, pp.13-18 and Paris, pp.23-4, and in the
zodiac facade at Chichen (cf. Kelley, 1976:47-50 and Fig. 4). In Cospi
they fall into three groups characterised by spike, watery shell, and
heart. This last, which includes the constellation Turtle or Gemini, is
shown to be good for hunting (pp.27-31), while the hook-tail constellations
of the first group represent a poisonous or hostile night sky (?Leo,
Scorpio; pp.21-4). In between watery Aquarius appears indifferent and
belongs to the only two women in the Cospi team of eleven, who are also
characterised by presiding over the only pages with even rather than odd
number bases (pp.25-6). Overall the page subtotals read:

Fig. 4: Divisions of the zodiac year, into 11 figures and 3
groups of emblems. Based chiefly on Cospi, pp.21-31 and
Fejérváry, pp.5-22.

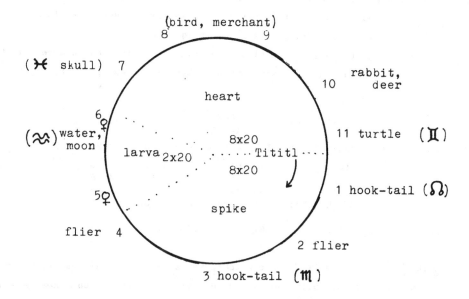

	spike male			shell female		heart male				
8 x	11	11	11	11	8	6	9	9	11	9	7
	297	341	189	189	72	54	2×9^2	9^2	11^2	9^2	7^2 (9)
	385	429	277	277	136	102	234	153	209	153	105

The total produced by this Cospi chapter, by means of simple addition, is 2460. That the astronomical reference here should be sidereal rather than synodic is suggested by the greater attention to the zodiac 'road' as such; and that it should include the moon is suggested by the divisibility of the total both by the lunar 9 and by a sidereal lunar period of $27.\dot{3}$ or $3(3^2 + \frac{1}{3}2)$ days. The actual loss per moon against this ritual figure is 0.012 days, or 3 in 2460 (90 such moons).

Recently, the much-neglected sidereal moon has been shown to underlie the Inca ceque system in Peru, in counterpoint to the obvious attention of that system to the synodic moon and the four quarters of the solar year (Zuidema 1981). That it had an analogous function in Mexico is indicated by the appearance in two iconographic texts of the total 246, i.e. 9 rather than 90 sidereal moons and exact to the day, as the property of metropolitan centres. The first is Tenochtitlan whose tribute-accounts list exactly 246 subject towns in the four quarters around its own metropolitan area, to west, south, east and north; in turn these relate to the synodic moonthrough their total of 29 head towns (cabeceras, Table 4) and to the solar year through their initial quarterly Fasts toponyms (Table 1) and through the overall total of 365 towns under the gobernadores of the metropolis and four quarters combined (Mendoza Codex, pp.22-55):

gobernadores		head- towns		all towns										(10)
Petlacalcatl	123	centre	9	13	10	26	16	26	7	10	7	9	=	124
Atotonilco	46	west	7	6	7	13	12	6	2	1			=	47
Tlachco	69	south	7	10	14	12	14	8	6	6			=	70
Chalco	82	east	7	6	22	11	11	3	22	8			=	83
Cuauhtochco	45	north	8	7	6	7	11	7	2	5	1		=	46
	365		29											246

The other example is Tepexic which governs 29 provincial head towns in its Annals, to east, south, north and west (Table 4) and which likewise, in Aubin ms. 20 top left, holds subject to itself 246 dots on maguey paper, arranged as follows:

$$
\begin{array}{ccc}
116 & 88 & \\
3 \times 7 & 3 \times 7 & = 246
\end{array}
\tag{11}
$$

While leaving little doubt about the connection between the
246/2460 total and the sidereal moon, these last examples suggest that it,
too, is a 'super-number', like the Fejérváry's 2914. For the sheer
arrangement of the dots under Tepexic in Aubin ms. 20 points incontestably
to Mercury, sidereal as well as synodic, through the sub-totals 88
(including one 'hollow' or absent dot), 116, and 42 which is one and half
times 28, the difference between 88 and 116. (The hollow dots in Fejérváry
pp.15-22, less the lunar 9 on p.17, also total 116 + 88: see (6) above.)
Hence:

$$
\begin{aligned}
246 &= 9 \text{ sidereal moons} \\
&= 116 + 88 + 3(\frac{116 - 88}{2}) \\
&= 3^5 + 3
\end{aligned}
\tag{12}
$$

And the fact that the total in question is registered on the page as dots
on a trimmed paper page points in turn to a link between that planet and
the very idea of writing and arithmetic, which is also implicit in its
possible hieroglyphic identity as Sign III (Fig. 1): cf. the esoteric
Maya term 'akab dzib', 'Night (III) or Mercury Writing'. (In this case
the Mam and Tzental term for Sign III, uotan, would coincide exactly with
the Old World Mercury or Wotan). Be that as it may, sidereal and synodic
Mercury can also be deciphered in the Cospi total of 2460. For there the
difference between 88 and 116 serves to multiply Mercury's actual sidereal
period of just under 88 days to within one day to 2460 (2461), the
dominant sub-total in the calculation being 88, which frames each one of
the opening 'spike' pages (pp.21-4, Table 5c). Then, for good measure,
the 2460 total may finally be made to include the other 'plunger' Venus,
if we invoke the zodiacal 11 and the principle of shifting between dimensions of time. So that bringing together the sidereal periods of the moon
and both inferior planets, in ritual form, it may be read as follows:

$$
\begin{aligned}
2460 &= 90 \text{ sidereal moons of } 27.\dot{3} \text{ or } 3(3^2 + \frac{1}{3}2) \text{ days} \tag{13} \\
&\simeq (116 - 88) \text{ sidereal Mercury periods of} < 88 \text{ days} \\
&= 11 \text{ sidereal Venus periods of } 224.7 \text{ days, less 11 days}
\end{aligned}
$$

However these periods were calculated, and whether or not they
imply knowledge of a heliocentric system (cf. Aveni, 1980, chapter 3),
there can be little doubt about the sidereal element in them. In the case
of both the moon and inferior planets, the totals could have been arrived
at, without the need for geometry, by the purely numerical formula:

$$\frac{1}{si} = \frac{1}{sy} + \frac{1}{y} \tag{14}$$

where, in days, si is the sidereal period of the moon or an inferior
planet, sy its synodic period, and y the solar year. At the same time,
in the case of the moon, this sidereal period can be measured with no
difficulty at all by direct observation against the constellations of the
zodiac. And of itself it establishes the important principle of there
being two kinds of movement proper to the bodies passing along the ecliptic, synodic and sidereal.

THE SOLAR AND THE SIDEREAL YEAR

In the hieroglyphic inscriptions, it is generally accepted
that the length of the solar year may be registered, with high accuracy,
by dates which establish the difference which accumulates in the Era
between it and the metric year of 365 days. Seventh century examples of
such dates, first decoded by Teeple, are discussed by Thompson (1971:317).
Further, the particular span over which one year accumulates between the
solar and metric year appears to be registered in the twice 1.18.5 tun
date at Palenque (Temple of the Cross); this is equivalent to 1508 solar
years to the nearest year, or 29 Rounds.

In the iconographic tradition, just this 29-Round period, here
referred to as the Solar Span, appears in its own right, in the Flames of
the Sun on the title page of Laud (p.46) for example, and serves as a
factor of Era dates counted out in several calendrical texts. The 1570 AD
date in Mexicanus (p.9) mentioned above consists of 158 years added to
1412 AD, or three such spans beginning with 1 Flint of the first Round
3112 BC (3 x 1508 or 4524 years): the multiplication of the 29 Rounds by
three is effected by three turns of St. Peter's key. In Aubin ms. 20 (top
right) the same three turns appear in a coil under the year 5 Flint 1416
AD, which date is confirmed as 12 Rounds (star-eyes) less 14 years distant
from the year 5 House 805 AD at Tepexic (top left; cf. Tepexic Annals, p.
32). Again, in Tilantongo Annals or Zouche obverse (pp.3-4), Round 58 or

a double Solar Span is distinguished by the mention of the Solar Span as
such (14 + 15 star-eyes) and of lean and canine Stones which both Aubin
ms. 20 and Tepexic Annals show refer to the mechanism of dropped leap
days. For the former includes one such lean Stone in a row of 34 (136
years, top right) while in the latter (pp.31-41), between Rounds 74 and 79
of the Era, two such Stones characterise the two leap days that need to be
dropped over 5 Rounds or 260 years (Fig. 5). Hence, at Round 58, the
point in the Era where two years have accumulated between the solar and
metric year, Tilantongo Annals alludes to the whole principle of the leap
days that distinguish the former from the latter.

       The double Solar Span of 58 Rounds or 3016 years proves like-
wise to be integral with the first number sequence in Fejérváry, pp.5-14.
For moving up to the page-register above the bars-and-dots, from days and
years to Rounds, the Round-Flames of the ceremonial fires that burn before
the male and digital members of the eleven total 58, which is also the
total of the Sticks that are their source (cf. Table 5a; only incense
smoke appears in Cospi). Hence as the modified 50-year Rounds or
xiuitlamolpia defined for us by Molina (here coloured red as exact eighths
of the Head - cf. also their use as such in Laud, p.44), the Sticks
relate to the Round-Flames through the 116 of Mercury, a total confirmed
in the 58 red dots on the knots that literally contain the 2-year difference between them:

       Fig. 5: Tlach-malacatl, equinoctial Court-and-spindle,
       Tepexic Annals, p.40. Shows leap day dropped from the six
       epagomenal days (ball, feet), and occurs twice over 5 Rounds
       before the seventh kindling, the first to be measured as an
       exact Round (7 Reed, 963 to 1015 AD). The same formula
       appears in Borgia, pp.49-52, where the two dropped days are
       juggler's balls.

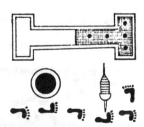

$$116 = 58(52 - 50)$$
$$= 2 \ (6 \ 6 \ 9 \ 9 \ 9 \ 7 \ 12) \ \text{red dots} \tag{15}$$

And it is these Sticks and dots which offer to relate the 58-Round or 3016-year span to the original 2914 cipher. For on p.13, where two of the Flames separate off uniquely to engulf the incense which in its turn uniquely gives off no odour, the 2-year dots, seven in all, are uniquely concentrated on the Sticks alone. Hence:

$$2914 = (58 \times 50) + (7 \times 2) = 58 \text{ Sticks plus } 7 \text{ dotted knots}$$
$$= 2 \text{ Solar Spans less } (2 \text{ Rounds less } 2 \text{ years}) \tag{16}$$
$$= 58 \text{ Flames less } (2 \text{ Flames less } 2 \text{ years})$$

The measurement of the solar year implied by the recognition of the 1508-year Solar Span is more accurate than that of the Gregorian Calendar:

$$\frac{365.2422}{1508} = 0.2422 \tag{17}$$

In the annal narratives, the 29-Round Solar Span, over which the difference between the metric and the solar year amounts to a year, finds a logical counterpart in the Sidereal Span, over which the difference between the metric and the sidereal year amounts to a year. In the Tepexic Annals, on the year and day 2 Reed 2 Reed 1681 BC, 27 Rounds and years in the Era and 1427 years from the first named year 5 Flint 3108 BC, we find a clear depiction of slippage from one star to the next relative to a fixed socket or crater below (Fig. 6). Over this period the metric year does in fact lose exactly a year against the star or sidereal year. Moreover the star image is recalled, with double slippage, at Round 55; a

Fig. 6: The Sidereal Span, Tepexic Annals, p.10. Shows slippage against the stars at 2 Reed - 2 Reed, Round 27, 1681 BC, 1427 years from the first named year 5 Flint, 3108 BC.

very close approximation to two such Sidereal Spans, between 5 Flint 3108 and 5 Flint 248 BC, this moment is here ritually rounded up to 5 Rounds, star-eyes, plus 5 decimal Rounds, an Eagle (Table 6b). The particular 2 Reed date also defines a trine series of 9 Rounds which culminates in the first kindling, after 27 Rounds, and runs through to the second and last kindlings, i.e. between Rounds 39, 48, 57, 66, 75 and 84 (pp.12, 13, 15, 21, 32, 48); this 2 Reed date appears nowhere else in the narrative and on its first appearance, at 1681 BC, indicates an astonishingly accurate sidereal year measurement of 365.256 days:

$$\frac{365.25636}{1427} = 0.25596 \tag{18}$$

Along with the Solar Span, the double Sidereal Span is invoked on the title page of Laud (p.46) by Hell-Lord at the nadir (5') and Eagle at the zenith ('5'), making 55 Rounds in all. And there appears to be another ritual expression of the single Span in Borgia, p.72 where Heroes 9, 5, 5, and 8, ranged in a swastika round the celestial spider (between Rounds 72 and 99), gesture toward xiuitl leaves that total 2(5 + 6), making 27 x 52 plus 22, or 1426 in all:

$$\frac{365.25636}{1426} = 0.25614 \tag{19}$$

Given the hieroglyphic evidence, there is no reason not to accept a highly accurate determination of the solar year in Mesoamerica. In the case of the sidereal year, while evidence remains undeciphered in hieroglyphic and is scant in iconographic script, there is no reason in principle why it should not have been determined with equal accuracy, by direct observation of the sky, and the gathering of a data-base commensurate in time with maize agriculture in Mesoamerica.

THE PRECESSION OF THE EQUINOXES

It is indeed a bold claim that Mesoamerican chronographers knew of the existence of the 26000-year precessional period. They recognised it as that cycle over which a zodiacal constellation regains the same position in the solar year, say heliacal rise occurring at the spring equinox. Yet two things encourage us to think they might have done. The first

stems from the categorical distinction established so far between synodic and sidereal periods
in Mesoamerican astronomy, and its possible application to the year. For it is precisely the
difference between the solar and the sidereal year which numerically produces the 26000 or
so years of the precessional cycle. Astronomically this difference is unstable after the second
decimal point and notoriously hard to predict; but at 0.014 of a day (365.256 less 365.242)
it yields the 26000 years or 500 Rounds of the cycle (26000 x 0.014 = 364 days), modern
estimates being some 25730 years.

This leads to the second, far stronger source of evidence.
For the length of the cycle, here called the 'great year', appears
measured out in its own right, like the far smaller Solar Span, in ritual
and calendrical texts. Indeed its presence in iconographic 'ritual'
texts is as much a diagnostic of that genre as is a chaptered format and
the practice of multiple reading. The time-markers involved include not
just Rounds but Rounds numbered by the 13 Heroes, and Decimal Rounds or
Cross-Eagle Counts of 520 years (10 x 52, 13 x 40) numbered by the 13
Quecholli; nor just Heads but the 2000-year Macuiltzontli with its special
Flower-Heart epithet (Table 3). But throughout, firm corroboration is
given by normal Round and Head markers in other texts and by certain
hieroglyphic datings.

Chronologically the great year serves to set the history of
our local era into cosmic time, the Era of 5200 years being both a fifth
of a great year and the fifth of the Suns of creation. Hence it typically
occurs in this cosmic context, along with the tonalamatl names of this Era
- 4 Quake, which is the spring equinox of 3113 BC - and of the Suns prior
to it, as well as with the introduction of the first years to be named by
the Series Signs of the tonalamatl.

On the Sunstone of Tenochtitlan (Table 6a), the 500-Round
total of the great year is emblazoned in the 10 x 10 cloud-Flames on the
snakes that hold this Era's Series III year 13 Reed 3113 BC between their
tails and encircle its Sun-Sign 4 Quake, itself composed of the Signs of
the four previous Suns; the one snake wears the mask of the Sun, Hero 4,
and the other the mask of the Fire-Lord Hero 1, making 400 + 100 in all.
Hero 4 also personifies the 400 pre-Era Rounds in Borgia, p.71 when he
sits on the throne of its Sun-Sign 4 Quake and holds his 100 in 5 Flags
blood-marked as Rounds. Again, in the Cuauhtitlan Annals, the same 400
Rounds appear as the 'CCCC' Cloud-snakes: these are killed by Obsidian
Butterfly, whose Ennead-Quecholli name esoterically gives the C11 Round-7:
base within this Era (2' + '7'). As for calculations in Heads, these

inhere in the very structure of the 52-column 'Thrones and Suns' chapter that opens Borgia, Vaticanus B and Cospi, where moreover the 26000-year unit basic to the cosmogony of the four Suns is registered in the first of them, in the Valentine image of the 2000-year Flower-Heart, comprising 5 x 5 Jades, traversed by the 13-year Reed or quarter Round (Fig. 7). For its part, in the nutrition sequence beginning with the 'sucklers' tree' Chichiualcuauitl and its liquid and feathered xiuitl of 400 years, glossed as such, Rios specifies the following totals of Heads: < 20 (15 plus 2 consumed and 3 forming), 10, 10; 12, 13 (= 65 x 400 = 26000). In this scheme, this and the previous era bear to each other the lunar Parrot:Quetzal ratio 13:12 while together at 10000 years they bear the Venusian-solar Cross Eagles ratio 5:8 to the preceding 16000 years.

　　In Borgia, the Fasts chapter (pp.29-46) opens by appending Series I years to a great year reduced to 25600 years; this is registered in bars as 10 (two bars) x 32 (teeth) x 80 years (the central jade, p.30). As 64 rather than 65 Heads, a figure closer to actual precession, this foreshortening could explain the 1692 AD date, prominent in the Tepexic Annals and the Anahuac tradition generally, that dispenses with the last of this Era's Heads. Achieving a similar effect, the snake calculations in Dresden (p.61) count from a point 3 pictuns 3 baktuns and 3 katuns or about 25000 years prior to the minus 13 baktun base used at Tikal (and analogous to the pre-Era minus 12 Head base in such texts as Rios, p.7). But for a really accurate estimate of precession we must return to Fejérváry.

Fig. 7: The Valentine image attached to the scheme of Suns and calculated as a 2000-year Flower-Heart (equal to 5 x 5 Jades) traversed by the 13-year Reed-arrow (2000 x 13 = 26000), Cospi, p.2.

Above the bars and dots of the octaeteris and the Flames of
the Solar Span in this work, at the highest level and page-register (Table
5a), there run two heavily-annotated sequences of tonalamatl dates (their
counterparts in Cospi are now partly illegible). In both cases the units
of the intervals they spell out are clearly qualified as Rounds (p.6), by
just the cloud-breath marker used for that purpose in Cuauhtinchan History
(yn quetzalteueyac, p.29), where it qualifies year- and day-names to produce the 59 + 13 + 10 Rounds in the Era prior to 9 Flint 1176 AD. Here,
the lower run of tonalamatl dates has Number dots that are double-
coloured and confirms the twice 29 Rounds already deciphered in the Flames
beneath (equation 16), adding to them the 500 Rounds of a great year (pp.
6-8):

$$558 = \quad 78 \ + 39 \quad + 130 \ + 182 \ + 129 \quad \text{Rounds} \tag{20}$$
$$(1 \ \text{II} \quad 1 \ \text{XX} \quad 1 \ \text{XIX} \quad 1 \ \text{IX} \quad 1 \ \text{IV} \quad 1 \ \text{I})$$

Then the higher run of dates (pp.9-12), whose thick-rimmed dots lack
colour, spells out the remarkable total of 990, ratified in Rounds by an
extra large Turkey ('9') that has a zero-enclosing 9 on its chest. This
is twice the previous 500 less 1%, a more accurate deduction than any
other known; and that two great years are involved is further suggested by
the fact that the Series III year-names that open and close the sequence,
7 House to 9 Reed, occur at the same juncture, as years of the 'Sun' and
'Age', in the Cuauhtitlan Annals and Mexitin History accounts of the Suns:

$$990 = \quad 97 \ + 23 + 201 + 193 \quad + \ 8 + 219 \quad + 78 \quad + 171 \ \text{Rounds}$$
$$(7 \ \text{III} \quad 13 \ \text{XX} \quad 9 \ \text{II} \quad 3 \ \text{IV} \quad 1 \ \text{XVII} \ 9 \ \text{V} \quad 7 \ \text{IV} \quad 7 \ \text{II} \quad 9 \ \text{XIII})$$
$$52 \times 990 = 25740 \times 2, \ 7 \ \text{House (III) to 9 Reed (XIII)} \tag{21}$$

An appropriate image for these two great years is provided by
the twin gold disks emerging from the treasure casket of solstitial
Etzalcualiztli (p.12), the quarterly Fast of metal tribute. For paired
with two further disks due from the casket after the next quarter they
indicate an overall tribute total of 4 + 4, the great octaeteris of over
200000 years shown by huge year markers on the Sunstone, which is finally
linked by that ritual cipher to the opening formula used with days and
years:

$$4 + 4 \begin{cases} \text{days of Venus inferior conjunction} \\ \text{years of the octaeteris} \\ \text{great years of the great octaeteris,} \approx 8 \times 500 \text{ Rounds} \end{cases} \quad (22)$$

The 4 + 4 underworld or winter-solstice passages of this great octaeteris are recorded in the Popol vuh (lines 1675-4708), and in Rios (p.2) where they immediately precede the double great year (p.3, also ratified in Rounds by a huge Turkey) and the single great year of 65 Heads (pp.4-8). In Madrid, the 213740 years related to the ascent of man (p.72; 1.7.2.2.12 tuns) likewise falls into 4 + 4 parts.

In this perspective, the great year of equinoctial precession emerges as the missing link between the local and political chronology of our Era and the vast evolutionary philosophy of time so vividly testified to in the Popol vuh. In its further reaches, this cosmogony stretches over aeons, the millions of years recorded in hieroglyphs at Quirigua, which however may always be related back to the great year through ciphered shifts between dimensions of time. From a practical point of view, a culture as ancient, numerate and chronologically sophisticated as that of Mesoamerica is more like than not to have detected and then measured the phenomenon of precession.

SUMMARY

In this exploration of astronomical norms in the iconographic tradition, the strongest evidence stems from two obviously numerical chapters in the Fejérváry and Cospi screenfolds. Taken together, and related to other ritual and calendrical texts in iconographic and hiero-glyphic script, these sources suggest a coherent system of astronomical knowledge centred on the movements of sun, moon and the inferior planets through the ecliptic, among which the lunar-solar cycles of 8 and 19 years predominate. Of particular consequence are the numerical totals which refer incontestably to the synodic and sidereal periods of the moon and Mercury, a distinction in principle which seems to be valid likewise for the solar and sidereal year. In any case, the precessional great year which arises numerically from this last appears to lie at the heart of the Mesoamerican cosmogony within which our 5200-year Era is situated, as a fifth and the fifth of five Suns.

Table 1: Divisions of the year

a. The eighteen Fasts and Feasts of 20 days (1. = lesser; g. = greater):

		Nahua	Emblem	Maya Highland	Lowland
Jun	1	Etzalcualiztli	steaming pot, rains	mat, base	mat (pop)
	2	1.Tecuilhuitl	(quincunx, litter on)	1. jaguar	red) sky
	3	g.Tecuilhuitl	(high, Thor's Maundy)	g. jaguar	black) bands
	4	1.Micailhuitl	(mummy, black knife)	vampire	vampire
	5	g.Micailhuitl	( )	knife	(zotz)
Sep	6	Ochpaniztli	woman's broom, Court and spindle	dog	dog (xul)
	7	1. Pach	atlatl	1. Pach	
	8	g. Pach	water-source	g. Pach	jade, water
	9	Quecholli	quecholli, red feather, reed-arrow (XIII), knapsack	quecholli (tzikin); black road	black ⌉
Dec	10	Panquetza-liztli	flag, four roads	white road	green ⌡
	11	Atemoztli	subsiding water	red road	white
	12	Tititl	comb	green road	red
	13	Izcalli		Izcal	
	14	Atlacahualo	tree over water	tree (pariche)	tree
Mar	15	Tlacaxipe-hualiztli	xipe skin, mask, hat (year marker)	Tlacaxipe	owl (muan)
	16	1. Tozoztli	maize tassel	1. (maize)	maize
	17	g. Tozoztli		g. (tassel)	
	18	Toxcatl			

Cf. Thompson, 1971:104-118; Nowotny, 1961:243; see also Borgia, pp.29-46;
Fejérváry, pp.4-22; Popɵl vuh lines 1675-1708; Rios, years 1519-21;
Guevea Maps. Quarters marked in the Tribute records of Tlapa,
Tlaquiltenango, Tenochtitlan (Atotonilco and Tlachco for Fasts 1 and 6).

b. The Fasts and the lunar-solar Quecholli:

Jun				Sep		Dec				Mar			

```
   Jun            Sep        Dec              Mar

  1 2 3  4 5      6      7 8  9 10  11 12  13 14     15      16 17    18
 '1' '2'  '3'  '4''5'  '6'  '7'    '8'    '9'   '10''11'   '12'  ('13')
```

See Aubin tonalamatl, pp.1-20; note March and Owl ('10') above.

Table 2:   The tonalamatl, Nahua version

a. Nine Figures or Ennead: yohualtecutin, night lords

1'	Fire or Year Lord	xiuhtecutli
2'	Obsidian	iztli
3'	Embryo Sun	piltzintecutli
4'	Maize God	cinteotl
5'	Hell Lord	mictlantecutli
6'	Jade Skirt	chalchiutlicue
7'	Cloth Goddess	tlazoteotl
8'	Hill Heart	tepeyollotli
9'	Rain	tlaloc x 29, less 1 = 260

b. Thirteen Numbers

Quecholli or Fliers   Heroes

'1'	Hummingbird	huitzilin	1	Fire Lord (cf. 1')
'2'	" (blue)	huitzilin	2	Earth Lord
'3'	Hawk	huactli	3	Jade Skirt (cf. 6')
'4'	Quail	zozoltin	4	Sun
'5'	Eagle	cuauhtli	5	Workers
'6'	Owl	chicuatli	6	Hell Lord (cf. 5')
'7'	Butterfly	papalotl	7	Flesh Lord
'8'	Eagle (striped)	cuauhtli	8	Rain (cf. 9')
'9'	Turkey	huexolotl	9	Quetzal Snake (quetzalcoatl)
'10'	Eared Owl	tecolotl	10	Mirror Smoke (tezcatlipoca)
'11'	Macaw	alotl	11	Night Lord
'12'	Quetzal	quetzal	12	Dawn-House Lord
'13'	Parrot	toznene	13	Two Lord x 20 = 260

As the two highest fliers, the Eagles '5' and his bald companion '8' cross their necks in a literal multiplication sign in order to mark out the 40-year Cross Eagle Cuauhquecholli in the annals of the 'eagle' towns Cuauhtinchan and Cuauhtitlan (e.g. between 1259 and 1299 AD); also the toponym for the town Cuauhquechollan itself multiplies 5 dots by the bald Eagle '8' (Mendoza, p.42). Being qualified by each of the Numbers in turn, like the equivalent 40-day period or double Fast, these 40-year units then build up into the Cross-Eagles Count, which is a decimal Round or 520 years in length (13 x 40 = 10 x 52; cf. Borgia, p.71). In these larger calculations of the year calendar whole Cross-Eagle Counts may be repre-sented by the Quecholli, since when fat or housed these may ritually bear lots of ten Rounds equal to their Number; e.g. Butterfly ('7') bears 70 x 52 years. Cf. Eagle in Table 6b.

c. Twenty Signs, by fourfold series

Series I	Alligator	VI	Death	XI	Monkey	XVI	Vulture
Series II	Wind	VII	Deer	XII	Tooth	XVII	Quake
Series III	House	VIII	Rabbit	XIII	Reed	XVIII	Flint
Series IV	Lizard	IX	Water	XIV	Jaguar	XIX	Rain
Series V	Snake	X	Dog	XV	Eagle	XX	Flower

x 13 = 260

Table 3: Markers for year multiples

a. Multiples

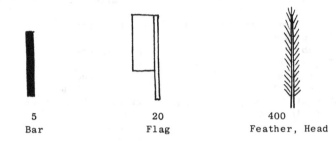

| 5 | 20 | 400 |
| Bar | Flag | Feather, Head |

b. Year Multiples

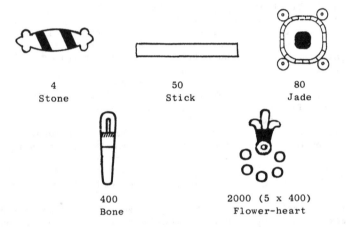

| 4 | 50 | 80 |
| Stone | Stick | Jade |

| 400 | 2000 (5 x 400) |
| Bone | Flower-heart |

50: as defined by Molina (1944) - 'xiuitlamolpia, termino de tiempo que
tenian y contauan de cincuenta en cincuenta y tres años'. (For the 53-
year period see Guevea Maps, Cuauhtinchan History, and Caso and Smith,
1966:30).
80: see Cuauhtinchan History, 1176-1256-1336 AD; Boturini, 1144-1224; and
Sigüenza, 1167-1247 ('Chalco').
400, 2000: cf. Laud, p.44.

c. The 52-year Round

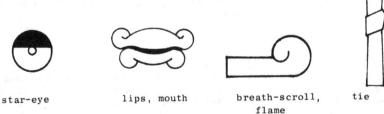

| star-eye | lips, mouth | breath-scroll, | tie |
| | | flame | |

In Cospi, pp.1-8 the tie is interchanged with the Rain god 9', also a
Round qualifier.

Table 4:   The tribute-field total of 29

	Headtowns	e	w	n	s	= 29	
a	Tepexic	7	5	12	5	Tepexic Annals, pp. 35 48 43 39	
b	Tilantongo	7	5	11	6	Zouche-Nuttall reverse, pp. 24 15 20 19	
c	Tenochtitlan	7	7	8	7	Mendoza, pp. 41 30 48 36	

	Quecholli						
d		12	5	8	4	Borgia, pp. 49-52	
e		12	1	3	13	Fejérváry, p.1	
f		12	1	3	13	Laud, pp. 31-38	
g		12	5	12		Palenque Trilogy panels 3 2 1	

	Other						
h		7	7	8	7	Mexicanus, p. 9 (cross, letters a-g)	
j		19	16	15	9	Dresden, pp. 25 27 28 26 (bar-and-dot)	
k		--	29	x	3	--	Tlaxcala Lienzo (squares)
l		7	7	8	7	Laud, p.46 (blood scrolls)	

a b e f have north as the upper region of 9'; a d e f have south as the
lower region of 5'.   a-c receive cocao from the east (in b west and south
should possibly be switched); c confirms the quarters by quarterly Fast
emblems and format change (cf. formula 10).   d-g place the quetzal in the
east; g makes north double as zenith and in the west Eagle is read from
warrior feathers (panel 2).   j equals twice the lunar 29.5 and bears the
moon glyph.   k appears under the four quarters and equals thrice 29, the
total to which h is also multiplied as the triple Solar Span in Rounds;
one such Span is recorded in l.

Table 5:   Specimen bar-and-dot pages

a. Fejérváry, p.12.   Below: 3 x 11 x 11 in bars and dots, sub-total of
the octaeteris of 2914 days.   Above: right, male figure, ninth in the
team of 11; centre, casket of quarterly metal tribute due at
Etzalcualiztli, with 2 + 2 gold disks; left, 8 Flames, 416 years, part of
the double Solar Span of 58 Flame-Rounds (or 3016 years; also 8 Sticks and
3 + 6 red dots, part of the 2914 year total).   Uppermost: four tonalamatl
dates with thick-rimmed Number dots, 9 V, 7 IV, 7 II, and 9 XIII (right
to left); conclusion of the span of 990 Rounds or 51480 years, confirmed
as two great years by the gold disks in the casket.

Table 5:   Specimen bar-and-dot pages

b. Fejérváry, p.15.   Main block and lower left margin: 3 x 10 (14 + 13 +
11) plus 10 (9 + 8 + 7 + 6 + 5) in bars and dots, part of the Micailhuitl
total of 2876 days and of the Metonic total of 6940 days.   Top and upper
left margin: symbols for serial addition and angle value, with the decimal
1 Dog (X; cf. Fig. 2).

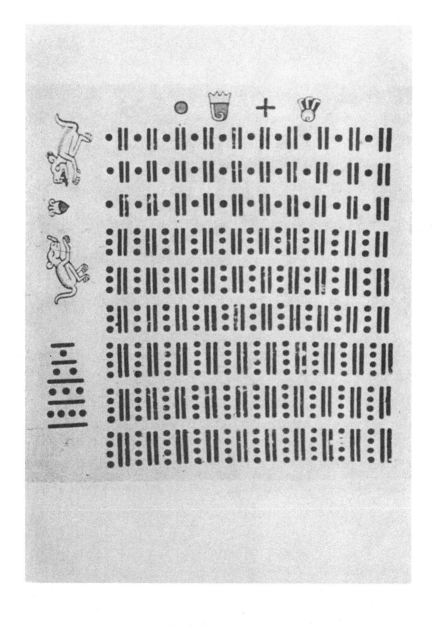

Table 5:   Specimen bar-and-dot pages

c. Cospi, p.21. Left: zodiacal emblems of the spike and stinger type.
Lower centre: 3 x 9 x 11 plus 4 x 2 x 11 in bars and dots, first sub-total
of the sidereal moon and inferior planet total of 2460 days. Upper centre:
armed male figure, first in the team of 11.

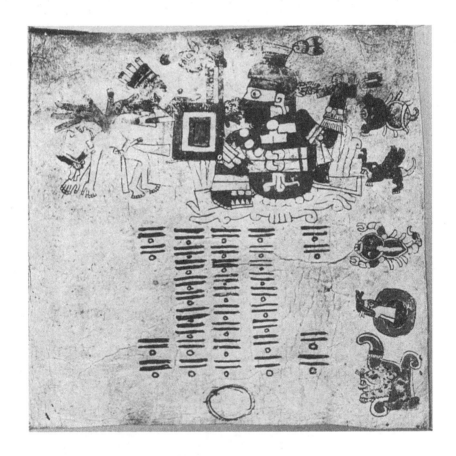

Table 6:   Specimen datings

a. Aztec Sunstone.   Top centre: inaugural year of the Era 13 Reed Cl,
3113 BC; possibly also 13 Reed Cll 1375 AD (cf. Mendoza, p.1) and 1479 AD
(assumed to be year of composition).   Right and left rim: 10 x 10 Round-
flames in the Hero ratio 4:1, i.e. 20800 pre-Era years right, and 5200 Era
years left; here the Chichimec Round 72 base 631 or 647 AD, appears as
4 x 18 ties.   Centre: the tonalamatl name of the Era or Sun 4 Quake
(XVII), incorporating the names of the four previous Suns.

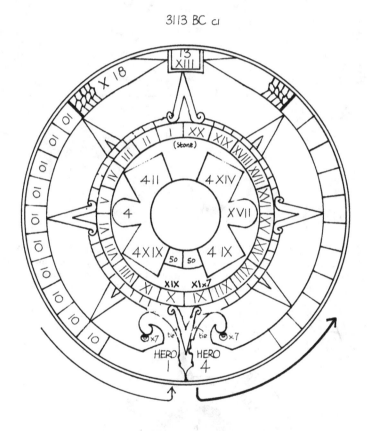

3113 BC cI

	BC	3113	
		3744	4 x 18 x 52 (ties)
	AD	631	13 Reed Cl
		16	Cl to Cll
		647	13 Reed Cll
		728	14 x 52 (eyes)
		1375	13 Reed Cll

Table 6:   Specimen datings

b. Tepexic Annals, p.14. Lower right: year 5 Flint 300 BC. Lower left:
year 5 Flint 248 BC, Round 55. Top: middle house of tribute confirms 5
Flint and the 5th year of the Round; the 55-Round distance from the Era
base is confirmed in the other two houses as 50 (Eagle) plus 5 (star-eyes).

BC 3113
2865   (5, 55 x 52)
BC 248 5 Flint Cl

c. Zouche-Nuttall reverse, p.35. Upper left: year 9 Reed C10, 1047 AD.
Lower right: a 'Jade-Round' or 80 x 52 years (the divisions of the Jade
amount to 5 9 7 5 6 7 7 6 or 52); 'nauhollin' means 4 Quake, the Era name
and an indication that the count is from the Era base.

BC 3113
4160   80 x 52
AD 1047 9 Reed C10

Table 6:   Specimen datings

d. Mexicanus, p.9.   Left (at 90°): Christian date reading 4 Heads less
3 Flags (shaded), 1540 AD, plus 18 dots (exclusive) to 2 Reed 1559 plus 12
(inclusive) to 1570.   Right: Era date reading '158' plus 29 (letters a-g
x 4 plus 1) x 3 (on St. Peter's key) x 52 (Series III Round) or 4682
years from 1 Flint 3112 BC, 1570 AD.

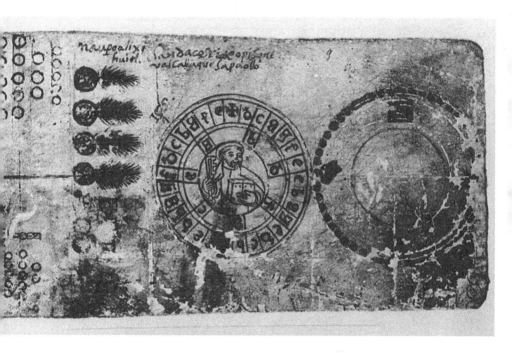

	BC 3112	1 Flint
	4524	3 x 29 x 52
AD	1412	
	158	'158'
	1570	

Table 6:   Specimen datings

e. Chichimec Maps i and ii.   1466 counted from 1 Flint Cl, Round 72, 632
AD (i); and from 1 Flint Cll 648 AD, the 'Chichimec base', 16 year places
or 4 Stones later and glossed 'yquhac oc centlelicue' to show this
difference (ii).

(i)

(ii)

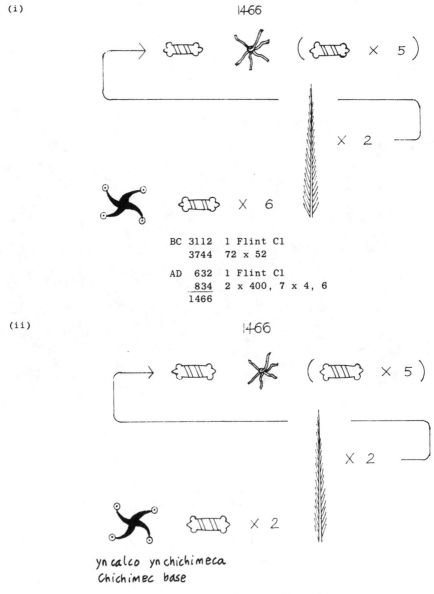

```
    BC 3112   1 Flint Cl
       3744   72 x 52

    AD  632   1 Flint Cl
        834   2 x 400, 7 x 4, 6
       1466
```

yn calco yn chichimeca
Chichimec base

```
      648   1 Flint Cll
      818   2 x 400, 3 x 4, 6
   AD 1466
```

REFERENCES

I Native texts

With Census number in Glass et al., 1975, which gives full bibliographical details; pagination follows revisions by Corona Núñez, 1964-67, except in the single case of Laud.

Aubin Ms. 20 (14)
Aubin tonalamatl (15)
Borgia (33)
Boturini (34)
Chichimec Maps (46)
Chumayel, Chilam Balam Book of (60)
Cospi (79)
Cuauhtinchan History (359, Historia Tolteca-Chichimeca)
Cuauhtitlan Annals (1033)
Customs of New Spain (188, Magliabecchi)
Dresden (113)
Fejérváry (118)
Guevea Maps (130)
Laud (185; pp.2-22, 23-46 for 46-25, 24-2)
Madrid (187)
Mendoza (196)
Mexicanus (207)
Mezitin History (1111, Legend of the Suns)
Palenque Trilogy of panels ('Cross', 'Sun', 'Foliated Cross')
Paris (247)
Popol vuh (1179; line numbers from Edmonson, 1971)
Rios (270)
Sigüenza (290)
Tepexic Annals (395; Vienna obverse)
Tilantongo Annals (240; Zouche-Nuttall obverse)
Tlapa Tribute-book (22; Azoyu 2 reverse)
Tlaquiltenango Tribute-record (343)
Tlaxcala Lienzo (350)
Vaticanus B (384)
Vienna (395)
Zouche-Nuttall (240)

Sunstone, Museo Nacional de Antropología, Mexico

II Secondary

Aveni, Anthony F. (1980). Skywatchers of Ancient Mexico. Austin and London.
      (1981). Archeoastronomy in the Maya region. Archeoastronomy, 3:S1-S16.
Brotherston, Gordon. (1976). Mesoamerican description of space: signs for direction. Ibero-Amerikanisches Archiv, 2:39-62.
      (1976 a). Time and script in Mesoamerica. Indiana (Berlin), 3:9-40.
      (1979). Image of the New World. The American Continent portrayed in Native Texts. London, New York.
      (1981). Das Kalendersystem der Maya und Mexikaner. Mexikon, III:24.

(1981 a). Year 13 Reed equals 3113 BC: a clue to Mesoamerican chronology. New Scholar, 8 (in press).

(1981 b). A Key to the Mesoamerican Reckoning of Time. London: British Museum Occasional Papers (in press).

Casas, J. Broda de (1969). The Mexican Calendar as Compared to Other Mesoamerican Systems. Vienna.

Caso, Alfonso. (1967). Los calendarios prehispánicos. Mexico: UNAM.

Caso, Alfonso & Smith, Mary Elizabeth. (1966). Interpretación del Códice Colombino. Mexico.

Clerke, Agnes Mary. (1911). The Zodiac. Encyclopaedia Britannica, 11th edition, 28:993-998.

Corona Núñez, José. (1964-67). Antigüedades de México basadas en la recopilación de Lord Kingsborough. Mexico, 4 vols.

Edmonson, Munro. (1971). The Book of Counsel: The Popol vuh of the Quiche Maya. Tulane.

Glass, John B., Gibson Charles & Nicholson, H. B. (1975). A census of native middle American manuscripts. In Handbook of Middle American Indians. Austin, 14:81-252, 253-280; 15:322-99.

Kelley, David H. (1976). Deciphering the Maya Script. Austin and London.

Molina, Fray Alonso de. (1944). Vocabulario en lengua castellana y mexicana (1571). Madrid.

Neugebauer, Otto. (1976). History of Ancient Mathematical Astronomy. Berlin, New York.

Nowotny, Karl. (1948). Erläuterungen zum Codex Vindobonensis (vorderseite). Archiv für Völkerkunde (Vienna), 3:156-200.

(1958). Die Ahau equation 584283. Miscellanea Paul Rivet dictata (Mexico), 1:609-34.

(1961). Tlacuilolli. Die mexikanischen Bilderhandschriften. Stil und Inhalt. Berlin.

Remington, J. A. (1981). Some thoughts about the Maya Long Reckonings. Paper delivered at the Oxford Archeoastronomy Symposium, 1981.

Seler, Eduard. (1901). Codex Fejérváry-Mayer. Eine altmexikanische Bilderhandschrift. Berlin.

(1902-4). Gesammelte Abhandlungen zur amerikanischen Sprach- und Alterthumskunde. Berlin, vols. 1 and 2.

Thompson, J. Eric S. (1971). Maya Hieroglyphic Writing. Norman (3rd ed.).

Zuidema, R. T. (1981). The astronomy and calendar of the Incas. Paper delivered at the Oxford Archeoastronomy Symposium, 1981.

ASTRONOMICAL KNOWLEDGE AND ITS USES AT BONAMPAK, MEXICO

F. G. Lounsbury
Yale University, New Haven, CT 06520, U.S.A.

"Needless to say, an unchallengeable correlation of the two
calendars [Mayan and European] would be immensely helpful
in identifying astronomical data in the texts, although I
myself am far from convinced that planetary observations
were recorded on stone monuments, unless favourable phenomena
perhaps governed a ruler's accession date."
—J. E. S. Thompson (1974: 85)

This paper is concerned with Maya attention to the planet
Venus, focusing on information that comes from one site, Bonampak, in the
state of Chiapas, southern Mexico, during the sixth through the eighth
centuries of our era. References are made to data from a few other sites
for purposes of comparison.

Six pages in the Maya hieroglyphic book known as the Dresden
Codex tell us most of what we have known about the Maya knowledge of Venus.
Five of these pages detail the subdivisions of the five synodic periods
that correspond approximately to eight solar years, the anciently known
eight-year cycle of Venus [5 x 584 = 8 x 365, = 2920]. The schema of
canonical periods for morning star, superior conjunction, evening star,
and inferior conjunction [236 + 90 + 250 + 8 = 584] is repeated thirteen
times on each of these five pages, bringing it into concord with the 260-
day divinatory almanac, and uniting these into a grand cycle of 65 synodic
periods of Venus, equal to 146 of the almanac and to two of the 52-year
rounds of the calendar [65 x 584 = 146 x 260, = 2 x 52 x 365]. The sixth
page, a preface to the other five, marks the historical institution of a
device that ingeniously accommodates this scheme to the need for correcting
accumulations of small errors (5.2 days in 65 Venus periods or 104 years),
accomplishing this without altering the fit of the Venus calendar to the
almanac, and thus obviating the otherwise inevitable need for total revision. (Transcriptions and analyses may be found in Teeple 1930, Thompson
1950 and 1972, and Lounsbury 1978.) A substantial sequence of observations

and record-keeping must have preceded the discovery of the basic cycles, and even more the invention of the corrective device. The 'explanatory hypotheses' that were posited by the Maya to account for these regularities may be surmised from the accompanying figures and hieroglyphs. These depict and name the several presiding deities, the five warrior-guises of the Morning Star, and some of his likeliest victims. It is known from Mexican sources, both historical and pictographic, that the heliacal rising of the planet following inferior conjunction was regarded as a time of especial danger.

From the Codex we learn also the hieroglyph of Venus, which occurs in two principal variants (Fig. 1 a,b). These appear to be in free variation, since their selection correlates in no consistent way with either astronomical or textual context. The glyph is composite. Its first component, prefixed or superfixed, is the sign that is otherwise attested as signifying the color 'red' and as being phonetically chac in Yucatec Maya. Whether the Maya saw Venus as red, or whether the sign was employed here as a rebus for chác meaning 'giant', is uncertain. Yucatec names for the planet Venus given in the 16th-century Dictionary of Motul are chac ek and noh ek, respectively 'red (or perhaps giant) star' and 'great star'. In other Mayan languages names that mean 'great star' predominate, though there are others, particularly for the Morning Star. If one of these was applicable to the hieroglyph (which in the Codex table stands for all aspects and phases of the planet) one is led to conclude that the second component of the glyph, in either of its principal forms, was literally and most simply 'star'. In one instance however, where its reference to Venus is certain (Fig. 1 c), and in five others where its reference is probably the same, the presumed 'star' component is employed by itself without the prefix. The point of this--and for occurrences of the glyph in the inscriptions it is a pertinent point--is that the main component by itself can be taken in either of two senses: either in a general sense as 'star', or in a specific sense as 'THE star' or 'Venus'. Further, it may be noted that this 'star' or 'Venus' sign serves occasionally in the inscriptions as an alternate sign for the day Lamat, whose more usual form is derivative.

There is yet another set of contexts in which this 'star' or 'Venus' sign appears, in which it has posed a curious puzzle. In these it stands as a superfix over an 'earth' sign (cab, Caban), or alternatively over a so-called 'shell' sign (of uncertain meaning and reading), or--as

yet another alternative--over one or another place-name sign, i.e., over
the main component of the 'emblem glyph' of a site, with an adjoined locative prefix. And in these it is usually flanked by a pair of like affixes
which are diagnostic though of unknown value. These glyphs (Fig. 2) had
once been suspected of having an astronomical meaning, possibly in relation to Venus, or possibly even to some other 'star'. But then some occurrences became known that put the matter in a quite different light and
led to another, nonastronomical, hypothesis. Two of these occurrences are
from a pair of inscriptions that tell the same story, in which the glyph
in question designates the first of a sequence of three events, the second
and third of which are the capture of an enemy king in battle on the second day and a standard ritual event (possibly a blood offering of some
sort) six days after that. On both of these monuments the bound captive
is shown under the feet of the victor, and on one of them his name and his
emblem glyph (his rank and his home address) are still legible. They
identify him as the king of Seibal. But it is precisely the Seibal site
name, with a prefixed locative, that stands under the 'star' or 'Venus'
sign in the first of the three event glyphs, the one in which we are interested. This seemed to force a conclusion that the 'star-over-Seibal'
glyph designates the raid, or some circumstance of the raid, on that enemy
capital; and that the associated date is the day of that raid. According-
ly, the glyphs of this category--'star over earth', 'star over shell',
'star over place-name'--have more recently been interpreted as having to
do with warfare; and this has turned out to be a tenable and productive
hypothesis. Yet, following the line of the earlier suspicion, David Kelley in exploratory papers (1973, 1977) was able to show that in some cases
--of sufficient number to warrant interest--the dates of such events could
be sorted into sets, within which they were separated from each other by
intervals that tended to approximate some multiple of 584, or they occurred on almanac days that tended to cluster around one or another of the
canonical days for a phenomenon of Venus in the Dresden Codex table. This,
then, has made it difficult to dispel the suspicion of an astronomical or
astronumerological significance for at least some of the 'star-over-X'
dates.

        With this as background, attention can be turned now to Bonam-
pak.

THE BONAMPAK MURALS

The site of Bonampak is in eastern lowland Chiapas, at 16° 44' north latitude and 91° 05' west longitude. Its principal known structure, a three-room building set on a low platform pyramid, with once richly painted interior walls and vaults, was discovered in 1946. The murals were photographed and rendered by artists within the next three years, during brief field trips under difficult circumstances. Further copies, both photographic and painted, were made in 1964. Few of the photographs, from either 1946-49 or 1964, have yet been made available. So it is the artists' renderings on which we must principally depend. The hieroglyphs, unfortunately, received something less than their proper due in these attempts at documentation, for neither with photographers nor with artists could they compete for attention against the rich and fantastic costuming in the ceremonial scenes of Rooms 1 and 3, or against the drama of the battle scene and the humiliation of prisoners in Room 2. And though the artists did an otherwise remarkable job of copying, they were not epigraphers. Nevertheless, the two dates in the hieroglyphic text of Room 1 have been ascertained with confidence; and the reconstruction of a third date, in Room 2, can be accepted I believe with equal assurance. As for what happened on those dates, at least the event associated with the third of them, in Room 2, is clear from the murals: it was a victorious raid on an enemy kingdom, resulting in the capture of prisoners. And as for those associated with the first two dates, it can suffice to refer to them in general terms as dynastic rituals; for the murals of Room 1 guarantee at least that much of a conclusion. If there was a date in Room 3, where the victory celebration is in progress, it has not survived.

Eric Thompson (in Ruppert et al. 1955) read the initial date of the text in Room 1 as 9.18.0.3.4 in the Maya day count (i.e., day no. 1,425,664), which is 10 Kan in the 260-day almanac, and 2 Kayab in the 365-day calendar year. In Thompson's correlation of the Mayan and European calendars, this would be A.D. 790, December 10 of the Julian calendar. The date was further characterized in the Maya manner for its attributes in the lunar calendar, namely moon age, position in the half-year, month duration (whether 29 or 30 days), and another detail not understood. It was the dark of the moon, precisely a date of lunar-solar conjunction. Three months had elapsed in the half-year, which at that time was reckoned in such a way that eclipse seasons came approximately in the middle of the half-year. It was not, however, an eclipse date; although it was just

five lunar months (148 days) after the only known instance of an apparent
solar eclipse date specifically designated as such on a Maya monument (cf.
Teeple 1930: 115).

The second date of the text is stated to be 16-score and 16
days after the initial date. It is 8 Ahau 13 Muan in the calendar round,
of which the 13 Muan is still legible but the 8 Ahau is not. The redun-
dancies in all this insure the accuracy of the interpretation, even where
parts are no longer legible. This date, by the same correlation, would be
A.D. 791, November 11. As already noted, the events on these two dates
appear to have been dynastic rituals of some considerable importance.

The third date is in a brief text accompanying the depiction
of the victorious warrior-king in Room 2, as he administers a ritual coup
to the principal captive at the culmination of that day's battle. The
numerical coefficients of the two components of the calendar-round date
are crystal clear: the day was 13-something 13-something. But the day
sign and the month sign are faded beyond recognition. Since the sequence
of the events depicted in the murals of the three rooms pretty surely corresponds to the numbering that has been assigned to these rooms, and since
there is no accompanying distance number or other anchor with this date,
it can be assumed to be the next '13-whatever 13-whatever' after the last
date in Room 1. Now this is a problem that allows of a fairly large number of solutions. The real problem is how to choose from among them, in
case there should be reason for wanting to know which one it actually was.
As it happens, there was someone who had such a reason and did want to
know; and that is how I got drawn into all this.

Early in September of 1980 Miss Mary Miller, then a graduate
student in Art History at Yale University who was writing a dissertation
on the murals of Bonampak, showed me this problem and inquired about the
possibility of a solution. She expressed the suspicion, moreover, that
this now hidden date might have had some astronomical significance for the
Maya. She had some interesting reasons for the suspicion, deriving from
her analysis of the murals where the date is found.

At the top of the sloping vault of the north wall of Room 2
(actually northeast), over the scene of the display of prisoners, but over
and opposite the scene of the battle, are four large cartouches containing
animal and human-like figures (see Fig. 3). They are above a 'sky band',
which position identifies them as representing celestial objects. The
leftmost is a pair of peccaries, with 'star' signs about them. The right-

most is a turtle, with three 'star' signs on its back. The two in between
are human-like figures, each with an adjacent 'star' sign. One holds a
star in his left hand and some other object in his right, while the other
partially supports himself with his left while apparently hurling a dart
or wielding a baton with his right. All of this is suggestive of zodiacal
astrology, for it is known from a thirteenth-century hieroglyphic codex
(the Paris Codex) and from the sixteenth-century Motul Dictionary, as well
as from twentieth-century native testimony (Thompson 1950: 116), that the
Maya have recognized a 'turtle' constellation. Further, a band of astral
or zodiacal signs inscribed on the Casa de las Monjas at Chichen Itza also
confirms this and gives evidence for a 'peccary' as well, with both the
turtle and the peccary depicted over 'star' signs within their respective
square cartouches.

It should be apparent now what it was that gave rise to Miss
Miller's suspicion and her inquiry about the possibilities for the partially obscured battle date. These 'star' figures and signs loom over the
scenes of battle and victory. She could not but wonder whether they, or
any one of them, might be an architectural and pictorial manifestation of
the concept that lay behind the 'star-over-earth/shell/placename' glyphs
(Fig. 2), which in a number of cases can be understood as designating
raids or battles, but which in some cases--as Kelley showed--seem yet to
have a connection with Venus calendrics. Her question was: Is it possible
that this Bonampak raid might have been timed for a Venus phenomenon? Or
for any other astronomical phenomenon? A related question was whether the
peccary and turtle constellations could be pertinent to the date of that
raid. And it was her suggestion that, whatever the calendar-round day
might be, it should be the next one of its kind after the dates of Room 1,
and probably at not too long an interval after the second of these.

To answer whether it was possible for the raid to have been
timed for a Venus or other astronomical phenomenon is of course easier
than to answer whether--assuming a positive answer as to possibility--it
actually was so timed. It requires only an enumeration of the possibilities and checking them out. But checking them out, with planetary tables
etc., requires a calendar correlation that can be applied with confidence.
A few years ago, in an article on Maya numeration and astronomy (1978),
I stated that the precise correlation between the Mayan and the Julian day
counts was still uncertain. That was a concession to the current state of
professional opinion on the subject, such as is appropriate in an encyclo-

paedia article, which that was. As for myself, I had little doubt then, and have even less now, that the Thompson correlation is correct. However, as I admitted, not all are agreed. So a new opportunity to test his or any other hypothetic values should be welcome. Here I shall restrict myself to Thompson's. Three different values of the correlation constant have been associated with his name: his original of 584285, his later revision (to accommodate Highland Guatemalan and Central Mexican data) of 584283, and the in-between value (proposed by Beyer) of 584284. It is clear now that for Classic Maya it is Thompson's original value, 584285, that is appropriate. It is that which I shall use here.

The dates recorded in Room 1, with Julian Calendar equivalents by the 584285 correlation, are as follows:

9.18. 0. 3. 4,   10 Kan 2 Kayab   (A.D. 790, December 10)
_____16.16   (recorded interval)
9.18. 1. 2. 0,   8 Ahau 13 Muan   (A.D. 791, November 11)

Miss Miller suggested that the 13... 13... date of the battle in Room 2 ought to follow these by a rather short interval. The possibilities for the next few years are the following:

9.18. 1.15. 5,   13 Chicchan 13 Yax   (792, August 2)
9.18. 2.13.10,   13 Oc 13 Mol   (793, June 23)
9.18. 3.11.15,   13 Men 13 Xul   (794, May 14)
9.18. 4.10. 0,   13 Ahau 13 Zotz   (795, April 4)
9.18. 5. 5. 0,   13 Ahau 13 Kayab   (795, December 20)
9.18. 5. 8. 5,   13 Chicchan 13 Uo   (796, February 22)
9.18. 7. 1.10,   13 Oc 13 Mac   (797, September 30)

Now it happens that the very first one of these is an astronomically interesting date, offering the possibility of a positive answer to Miss Miller's question, while the remaining dates are of no apparent interest. August 2 (Julian) of the year 792 was exactly the date of an inferior conjunction of Venus.

This is not all. At the latitude of Bonampak (16° 44') this date, which would be August 6 in retroactive Gregorian for that year, was also precisely the date of the sun's zenith passage, returning to the side of the south after having spent 89 or 90 days on the north side of the zenith. And with the ecliptic at 90° to the horizon on this date, it is of some interest that this inferior conjunction of Venus was one of those when the planet was at just about maximum celestial latitude (8° to the south, according to the Tuckerman tables); which, if I am not mistaken, means that the period of invisibility of Venus before and after the conjunction would have been just about minimum, perhaps as little as a day either side.

There is yet more in this date. At this conjunction Venus and
the sun were at the celestial longitude of 133° 35'. At the same time
(A.D. 792) the celestial longitude of Regulus was 133° 05', while its latitude was 0° 25'. The three heavenly bodies were in conjunction on that
day. It is worth taking a moment to picture what the Bonampak sky watchers would have been seeing (weather permitting) during the period leading
up to this date. A month and a half earlier (44 days to be exact) Venus,
approaching from the west, had come into conjunction with Regulus. It was
a close encounter, for they had only about a quarter of a degree of latitude between them at that time. Venus moved on, reaching its first stationary point 24 days later, having gone about eight and a half degrees
beyond Regulus; at which point it went into retrograde and came back to
Regulus, reaching it as both went into conjunction with the sun. At their
first encounter they were 42½ degrees (nearly three hours) east of the
sun; at the time of Venus's turnaround its elongation was 27 degrees (a
little less than two hours); and now at the second encounter they went
down with the sun, to emerge on the other side into the morning sky. We
know that the Maya knew enough to understand (in their terms) where the
planet and the star were at this time, even though temporarily lost to
view in the brightness of the sun.

Such was the date of August 2 of the year 792, which, if the
Thompson correlation is correct, was 13 Chicchan 13 Yax, 9.18.1.15.5. But
while the 13's in this Maya date are verified, being clearly legible, the
Chicchan and the Yax are hypothetic--merely one possibility out of several
that would satisfy the strictly calendrical requirements. On what grounds
then can we assume that this actually was the 13... 13... date that was
recorded for the battle scene of Room 2?

There are some tests, the outcomes of which will render the
hypothesis either more, or less, plausible. One of them is this. The
Chicchan-Yax combination was entertained because, in addition to being the
first in the set of possibilities, it was the only one that yielded an
'astronomically interesting' date. Now if that quality should render the
date unique in this respect among those recorded in the murals and in the
inscriptions at Bonampak, the date would be exceptional, without precedent,
and in no way expectable for the site. It could then be considered as
possibly no more than a chance coincidence, however intriguing. If, on
the other hand, it should turn out that it is not unique in this respect
among the dates recorded at Bonampak, if it should in fact be that the

majority of the dates recorded there are astronomically interesting ones, then it would be in conformity with the others, expectable, and rather more likely the correct interpretation of the evidence. And any other of the calendrical possibilities for the 13... 13... date would then be in some measure exceptional and less likely. Mary Miller's suspicion about this problematic date, and the apparent possibility of a positive answer to her question, prompted some testing of other dates at Bonampak. The results came as a surprise. They are of interest in their own right, and together they justify confidence in the 13 Chicchan 13 Yax solution of the battle date, which is at home among them, and which may be considered to have 'passed' this test.

The second test relates to the turtle and the peccaries, whether these constellations (if identifiable) were in any way pertinent to the hypothetic date of the battle, and thus whether they could be seen as significant components in its depiction. This matter will be discussed later; but in anticipation it may be noted here that, to the extent identifiable, they are appropriate and significant.

A third test relates to the question posed by Miss Miller, whether the placement of these 'star' figures over the battle and victory scenes of the murals might be considered an architectural and artistic rendering of the concept whose hieroglyphic expressions were the 'star-over-earth', 'star-over-shell', and 'star-over-placename' glyphs. This prompted testing of the date associated with the 'star-over-Seibal' and some of the other glyphs of this category. Only a beginning has been made and the results so far are mixed. About half of those tested have clearly interpretable and significant results in terms of critical points in the Venus cycle, while the remainder are of uncertain significance, relating possibly to a more general Venus criterion, or in some cases perhaps indicating no more than that there was a Venus-derived linguistic and hieroglyphic idiom for warfare. In terms of probabilities, the basis for the idiom in notions about Venus, particularly in relation to critical points in the planet's movements, is surely borne out by the results so far obtained. Eventually the matter will have to be submitted to a proper statistical evaluation; but before that can be done the examples will have to be collected and studied in relation to their hieroglyphic textual contexts, and their dates will have to be checked for astronomical attributes. A few of those already studied, which furnish comparative evidence pertinent to the Bonampak problem, will be reported here.

APPEARANCES OF THE EVENING STAR

It was the 'star-over-Seibal' glyphs on Aguateca Stela 2 and
Dos Pilas Stela 16 (Graham 1967) that gave the first clear evidence for
the military meaning which glyphs of this category are now known to have
had. The two monuments record the same events: a 'star' event on 8 Kan
17 Muan, 9.15.4.6.4 (apparently the date of the raid); a decisive battle
with capture of the local ruler on the next day; and a follow-up event
six days later. The hieroglyph of the first event suggests a Venus significance for that day. Its date was included by Kelley (1977) in the set
of those that cluster around the 16.6 positions in the Dresden Codex table
(the canonical days for beginning the evening-star period). If such a
table had general currency among the Maya of this period, and if it was
being followed to the letter, or numerologically, then the dates associated with a particular critical point in the Venus cycle might be expected
to conform, right to the day, to those of the table. The fact that they
do not, but that they depart in varying small amounts from them, suggests
that they were not determined solely by the calendar, but were more likely
dependent on observations. The date of the 'star-over-Seibal' event, by
the 584285 correlation, was November 29 (Julian), A.D. 735. According to
the Tuckerman tables, superior conjunction of Venus was 30 days earlier,
on October 30. An interval of 30 days after superior conjunction is a
reasonable one in the tropics, where the ecliptic is high and nightfall is
quick, for a first sighting of Venus as evening star (seen but briefly
after sunset, before it too sets--about 28 minutes after the sun in this
case). The tables indicate an elongation of 7.17 degrees for Venus on
November 29 of that year. Under good observing conditions in those latitudes the minimum necessary elongation for a first sighting appears to
vary from about five to about ten degrees, depending on the time of the
year and on the planet's celestial latitude at the time. (Similar limits
obtain also for a first sighting of the morning star at its so-called
heliacal rising, though the time required for attainment is much less.)
Thus, the position of Venus on this date was probably appropriate for a
first visibility of the evening star. That the circumstance should have
held significance for the Maya is inferrable in part from the glyph that
marks the date in these two inscriptions, and in part from the similar
situation of Venus on some of the star-marked dates at other sites. Two
of the Bonampak dates also catch Venus at this point in its cycle, though
they are without glyphic 'star' markings.

In Table 1 are listed some dates on which the eastern elongation of Venus falls within the indicated limits, and which are marked by
one of the 'star-over-X' glyphs, or by an unambiguous Venus glyph that
includes the distinguishing prefix (as in Fig. 1 a,b), or by a skull with
distinctive markings and teeth that is also a Venus symbol (Fig. 1 d), or
by this in company with one of the regular 'star' signs. Also the two
Bonampak dates that fall within this category are included. The dates are
listed and numbered in chronological sequence, but will be discussed here
in a different order. The columns of the table give the following information: (a) the day number in Maya notation; (b) the interval, in multiples of 583.92 [symbolized by the letter 'v'] plus or minus some number
of whole days, between any given date and a reference date which is indicated in this column with two asterisks; (c) the equivalent Julian date
by the 584285 correlation; (d) the celestial longitude of the sun for the
day in question [ca. 10:00 a.m. local time]; (e) the celestial longitude
of Venus at the same time; (f) the eastern elongation of Venus; and (g)
the interval in days after superior conjunction.

Date 4 in this list is the 'star-over-Seibal' date. It is
recorded at two sites in the southern Peten of Guatemala (those noted
above) which were under the dominion of some branch of the Tikal dynasty.
In the inscription of the Dos Pilas monument the event is denoted by the
hieroglyph illustrated here (Fig. 1 c). In that of Aguateca it is denoted
by a similar one, which compounds the 'star-over-shell' glyph with the
main component of the Seibal emblem glyph and a superfixed locative sign.
These glyphs, whose connotation seems to relate to Venus, apparently
designated the raid or a distinctive attendant circumstance. The engage-
ment of forces took place the next day, and the king of Seibal was taken
prisoner. He is so depicted, deprived of most of his finery and bound
with ropes, but identified by his name and emblem glyph, under the feet

Table 1.   First of Evening Star.

	(a)	(b)	(c)	(d)	(e)	(f)	(g)
(1)	9. 9.18.16. 3	100v +5	631 Dec 24	275.69	280.76	5.07	22
(2)	9.11. 0. 0. 0	87v −1	652 Oct 11	201.44	209.31	7.87	31
(3)	9.14. 0. 0. 0	50v +4	711 Dec 1	252.81	258.61	5.80	24
(4)	9.15. 4. 6. 4	35v −1	735 Nov 29	250.94	258.11	7.17	30
(5)	9.15. 9. 3.14	32v +1	740 Sep 13	174.20	181.72	7.52	28
(6)	9.15.15.12.16	28v −6	747 Feb 11	326.33	343.33	7.00	29
(7)	9.18. 1. 2. 0	* *	791 Nov 11	233.02	240.22	7.20	29
(8)	9.18. 9. 4. 4	5v +4	799 Nov 13	235.10	243.40	8.30	34

of his captor on both monuments. Six days later was the follow-up event
that has been mentioned above. Its nature is far from clear, although
there is quite a bit that can be said about it with fair assurance. What
is pertinent here is that it was apparently not necessarily lethal. Though
the conqueror took over at Seibal, the captive king was kept alive for
nearly another twelve years. He came to his end eventually as a sacrificial victim at a ritual ballgame--timed for an inferior conjunction of
Venus! (Details of this will follow under the appropriate heading further
on.)

Date 1, its event, and its designating hieroglyphs are an
earlier instance of the same sort as those of Date 4 which have just been
reviewed. Curiously, the date is just one day earlier in the calendar-
round, being on 7 Akbal 16 Muan. (Date 4, it will be recalled, was on
8 Kan 17 Muan.) The interval between them is equal to two calendar-rounds
and one day, or to 65 uncorrected Venus periods and a day, or 104 Maya
calendar years and a day; which would be just a day more than one complete
run of the Dresden Codex table. And like Date 4, this one also is recorded in two places, on Stela 3 at Caracol (Belize) and on the Hieroglyphic
Stairway of Naranjo (Peten of Guatemala). The records are of an engage-
ment between Caracol and Naranjo, the outcome of which was the conquest of
Naranjo by the ruler of Caracol, who then held power in both places. In
the Caracol inscription the event of this date is designated by the 'star-
over-shell' glyph compounded with the main component of the Naranjo emblem
glyph and a locative superfix, while in the Naranjo inscription (presumably promulgated by the victor after taking over) it is designated by a
'star-over-Naranjo' glyph without the shell. Even the glyphic variation
is parallel to that of the previous case. Astronomically, while the previous case represented a quite typical position of Venus at this point in
its cycle, the present one is about as close to the inside limit as is
possible, with an eastern elongation of little more than five degrees, and
a time interval after superior conjunction of only 22 days. It is an extreme case, and one wonders whether the Caracol war chief in his anticipation of the appearance of the evening star may not have jumped the gun
on it.

Date 6 is from the right-hand panel of the east doorway of
Temple 11 at Copan (Honduras). It is noteworthy for its association with
a complete Venus hieroglyph, which includes the distinctive <u>chac</u> prefix
as well as the principal 'star' component. This renders its primary sig-

nification unambiguous; it definitely pertains (in some way) to Venus. The recorded day, which is 5 Cib 9 Pop, was included in Kelley's list (1975) for this category because it approximates to within two days the 3 Ix 7 Pop position for the beginning of one of the evening star periods in the Dresden Codex table. If this is justified (and I believe that it is), then it requires the chronological position assigned to it here. It is one of the three earlier dates recorded in the temple from which it comes, and there are reasons for considering it to be the date of the ruler's designation as successor or acting regent in lieu of his prede-cessor who was being held captive at Quirigua. His accession to full title did not take place until seventeen years later (very likely because his predecessor, the legitimate holder of the title, was still alive). The event glyph in the passage with which we are concerned here has its two principal elements in common with those that accompany the Venus hieroglyph in each of the columns in the Dresden Codex table, which give the canonical days for the beginnings and endings of the morning and eve-ning star periods. The passage could thus easily be taken for a strictly astronomical record. But the glyph is also similar--even more similar because of a distinctive affix--to one that is employed in inscriptions to designate acts of heir designation, which glyph also employs the same principal elements. Moreover, the Venus hieroglyph, complete with its chac prefix, is employed elsewhere at Copan as a component in the string of appellatives and titles of the above-mentioned ruler. This usage, together with other elements in the present context, including its being preceded here by a 'lordship' title, suggest a literally applicable double meaning for the inscription: it was a date of the appearance of Venus as evening star, and it was the date of the designation of this ruler as act-ing regent and eventual successor to the captive king. (At least this seems to be the best hypothesis at the present time.) The Venus symbols and skulls that appear as iconographic elements in some of the monuments for which he was responsible may be indicative of the image that he ac-quired or had confirmed on this day. Astronomically the date is a typical one for this category, with an eastern elongation of seven degrees and an interval of 29 days after superior conjunction.

Date 8 is from what is probably the last hieroglyphic record produced at the site of Palenque (Chiapas, Mexico), a ceramic piece known as the Initial-Series Vase because of the fully specified date with which its covering text begins. It is the date of the inauguration of the last known ruler at that site. His six-glyph nominal and appellative phrase includes two components that have Venus associations, one of which is the 'star-over-earth' glyph, which also has connotations of warfare. Whether he carried this title solely by virtue of his accession date, or whether he had already acquired a reputation which his accession date symbolizes, is not known. The date is slightly on the generous side as a member of this set, with an eastern elongation of 8.3 degrees and a 34-day interval after superior conjunction.

Date 2 is a katun ending, from the middle panel of the Temple of Inscriptions at Palenque. As a katun ending it would have been recorded anyway, together with a declaration of the rituals performed for the occasion. At Palenque, during this period, other items of note were sometimes included. In the passage for this one a reference to Venus is indicated by a distinctively marked skull (Fig. 1 d) which is also a symbol of Venus. Since katun endings are determined by numerical criteria unrelated to Venus reckoning, the occurrence of a Venus phenomenon on this date is a pure coincidence. But it was one that did not escape the notice of the priests and the ruler of the place. What they did about it, other than include it in the record as a noteworthy feature of the katun ending, will not be known until the remaining hieroglyphs of that passage are better understood. Astronomically it is typical for the set, with an eastern elongation of 7.87 degrees and an interval of 31 days after superior conjunction.

Date 3 is another katun ending, three katuns later. A Venus phenomenon is implied for it in the record of Stela C at Copan (Honduras), and some sort of association with Venus seems to be implied by the iconography of Stela 16 at Tikal (Peten, Guatemala) which also commemorates this date. On Copan Stela C the evidence is indirect. A Venus event is twice imputed--in parallel passages in a couplet arrangement--to a mythological and numerologically reckoned antecedent date more than four and a half millenia earlier. In the first reference it is designated by a compound containing the usual Venus 'star' sign (as in Fig. 1 c), and it is preceded by an appropriate event glyph. In the second it is designated by the Venus skull sign (as in Fig. 1 d), this time with verbal affixes. The ascription to a mythological antecedent, in Maya practice, implies ascription of a similar or related attribute to the current date with which the mythological one is paired. But the three glyph blocks where this might have been recorded are destroyed; so direct confirmation is lacking. The

theme recurs, however, on the other side of the stela, where the same kind
of event is ascribed to yet another mythological antecedent. The three-
katun interval between Dates 2 and 3 is five days short of an integral
multiple of the Venus period [3 x 7200 = 37 x 583.92, -5.04]. Date 3 was
24 days after superior conjunction, and the eastern elongation of Venus
was 5.8 degrees. It is another one near the inside limit. But the date
was determined by a numerical criterion; and the first appearance of the
evening star, if not actual, was imminent. On the Tikal stela pertaining
to the same katun ending the text is brief. It records the day, that it
was the completion of fourteen katuns in the baktun, and that the required
ritual was carried out by the ruler, whose name and titles follow. The
text is disposed in three framed areas around the larger depiction of the
ruler himself, in regal attire, holding the emblem of office, but wearing
a headpiece of which the central component is the Venus skull backed by
the Venus 'star' sign. With the Copan and Palenque precedents in mind,
it might be supposed that the headpiece was motivated by the same consideration that is assumed to lie behind the manifestations of Venus significance in those monuments. It must be noted, however, that the same sort
of headpiece turns up again a generation later at Tikal, in Lintel 3 of
Temple IV. One would look, then, for a Venus attribute in one of the four
dates recorded in the text of that lintel. None of them, however, are at
any one of the usual critical points in the synodic cycle of Venus. The
event glyph of one of them is the 'star-over-shell' glyph (as in Fig. 2 b).
The date is 9.15.12.2.2, which by the correlation assumed here was July 28
of the year 743. On that date Venus as morning star (western elongation
23.63 degrees) was in precise conjunction with Mercury, with 0.64 degrees
of latitude difference between them. Whether this provides an adequate
explanation for the presence of the headpiece is not certain. But the
headpiece is at least indicative of an astrological concern with Venus
and some preoccupation with the power attributed to it.

        Date 5 is from Bonampak, from Lintel 3. The lintel depicts a
local ruler in one of the standard poses for administering a coup to a
captive. Though carved in stone and differing in details, in its style
and general outlines it corresponds to a focal scene in the battle murals
of Room 2, which scene depicts the ruler of a generation (or possibly two)
later in a similar pose, and contains the brief text with the 13... 13...
date that started this investigation. The victorious ruler on the lintel
wears a pendant jade skull mask hung from his neck, with the prominent row

of even teeth that characterizes the Venus skull, of which it is quite
surely a representation. The date of this event is a typical one of the
set of first appearances of the evening star, with an eastern elongation
of 7.52 degrees and an interval of 28 days after superior conjunction.
Since both the date and the pendant mask appear to implicate Venus in the
event of this lintel, they offer a supporting precedent for the reconstruction that was proposed for the missing components in the 13... 13...
date of the analogous scene in the mural, which also implicated Venus and
implied other astrological interests as well.

Date 7, also from Bonampak, is the second of those in the
mural text of Room 1. As noted earlier, it is the date of a dynastic
ritual of some sort (one suggestion that has been put forward is that it
may concern the formal presentation of a young heir to the throne.) That
it was a very important occasion can be judged from the costuming and
activities depicted in the mural scenes. The date is another one typical
for this set. The eastern elongation of Venus is 7.2 degrees, and the
date is 29 days after superior conjunction. It is a third instance of an
apparent astrological concern with that planet in the timing of important
undertakings at Bonampak.

MAXIMUM EASTERN ELONGATION

It was noted earlier that if it should happen to be the case
that the majority of the dates of planned events at Bonampak were astronomically interesting ones, then that would impart an additional measure
of plausibility to the reconstructed battle date; for it would then be
one of a kind. The two additional Bonampak dates which have been reviewed
above appear to do just that. The case would of course be strengthened
if there should be others. A check of the dates on the stelae of Bonampak
brings up the possibility of another significant category, that of the
maximum eastern elongation of the evening star. It is one that is supported by a few 'star' dates from other sites as well. Those that have
been looked into carefully are listed in Table 2, and are reviewed below.
The listings in the table are divided between two categories. The first
(category A) contains dates on which Venus is literally at its maximum
eastern elongation, having arrived at its goal, so to speak, in its journey away from the sun. For convenience, this will be called the 'arrival'
category. The second (category B) contains dates on which Venus has made
what may perhaps be interpreted as the first perceptible movement toward

a departure from that position, initiating the return journey back to the
place of the sun. In these the eastern elongations range from a quarter
to three-quarters of a degree less than those that were the corresponding
maxima. For naked-eye astronomy, with relatively simple sighting instruments, these may perhaps have been just-noticeable differences, enough to
indicate that the return trip was beginning. This will be called the
'departure' category. The listings of the dates are by chronological
order, but separately within the two categories. Their review, however,
will again take them up in an order different from that of their listing.
The numbering is continuous, proceeding from that of Table 1, rather than
beginning over again. The columns, except for the last, give the same
categories of information as do those of Table 1, namely: (a) the Maya
day number; (b) the interval to or from the chosen reference date, which
is that with the asterisks; (c) the Julian date; (d) the celestial longitude of the sun; (e) the celestial longitude of Venus; and (f) the
eastern elongation of Venus; while the last column (g) this time gives the
number of days before the next inferior conjunction. In part B of the
table, the bracketed figures in columns (f) and (g) give for comparison
the corresponding maximum eastern elongations and the intervals from their
dates to inferior conjunction. The bracketed figures thus represent positions of the same kind as those of part A of the table, and are those
away from which the presumed first perceptible moves of the planet have
been made. The amount of the movement may be seen by subtracting any
given unbracketed figure in column (f) from the bracketed one just below
it; and the number of days from the date of precise maximum to the recorded date can be seen from a similar comparison of the figures in column (g).

Table 2. Maximum Eastern Elongation.

	(a)	(b)	(c)	(d)	(e)	(f)	(g)
A. Arrival:							
(9)	9.12. 0. 0. 0	* *	672 Jun 28	99.12	144.72	45.60	71
(10)	9.12.11. 6. 8	7v +1	683 Sep 7	167.13	213.75	46.62	72
B. Departure:							
(11)	9.17. 5. 8. 9	* *	776 Jun 11	83.74	128.31	44.57	57
						[45.31]	[71]
(12)	9.17.10. 6. 1	3v +0	781 Mar 29	12.50	57.52	45.02	58
						[45.68]	[70]
(13)	9.17.15. 3.13	6v +0	786 Jan 14	298.66	345.10	46.44	60
						[46.99]	[73]

Three stelae are known from Bonampak, on which there are
recorded a total of five dates. Two of these are what Mayanists call
hotun-ending dates, marking the ends of round-numbered five-tun intervals
in the day count (the Maya lustra, quarter-katuns, = 5 x 360 days), which
were the occasions for erecting commemorative stelae at many sites. Dates
of this category are determined by strictly numerical criteria, and so may
be excluded from consideration in connection with astronomical matters,
except when such a date happens to coincide with an astronomical phenomenon of interest, of which note is made in the hieroglyphic text. Two
examples that did merit such notations, which were of katun endings, were
seen in dates 2 and 3 (Table 1); and another, yet to be discussed, is
date 9 (Table 2).

Stela 1 at Bonampak commemorates the hotun ending of 9.17.10.-
0.0, recording the date and reporting the enactment of the required rites
by the then ruler, whose name, parentage, and titles follow, and whose
standing full-figure portrait is the focus of the monument. The date is
not one of any special astronomical interest, and no such notation is
included in the text. The next stela, chronologically, is the one that
is archaeologically designated as Stela 3. It commemorates the hotun ending of 9.17.15.0.0, recording the date, the ritual, and the name of the
same officiating ruler, and depicting him facing a kneeling bound captive.
Again the hotun ending is of no particular astronomical interest, and the
inscription contains no such reference. But there is a subsidiary passage
in which the record of the event is almost totally obliterated due to
erosion, but in which there are clues to the date. The passage begins
with a one-block (and therefore two-digit) 'distance number' that is mostly lost due to breakage; but it is apparent that the number of uinals
(units of 20 days) was either 3 or 4, while there is little secure evidence as to the number of days beyond that. (See Robertson 1980 for a
photographic plate, and Mathews 1980 for a drawing.) Following that there
is a clear 'posterior date indicator', and then a calendar-round day that
has been interpreted in the drawing as 12-something 16 Cumku, the identity
of the day sign being unclear. A check of all possibilities, however,
shows that there is no two-digit distance number that can lead from 9.17.-
15.0.0, which is 5 Ahau 3 Muan, to a day 12-something 16 Cumku. It is a
mathematical impossibility. But a distance number of 3.13 will lead to
13 Ben 16 Cumku. The conclusion that must be drawn is that the coeffici-
ent of the day sign (with two dots and an intermediate space filler) has

been misinterpreted (the space filler needs to be another dot), and that the originally recorded distance number was 3.13 and the posterior date was 9.17.15.3.13, 13 Ben 16 Cumku. Reexamination of the photograph shows that it permits of this interpretation.

This is date 13 in Table 2. As indicated there, the eastern elongation of Venus on that day was 46.44 degrees, which is very close to the maximum attainable; and the date was 60 days before the next inferior conjunction. Precise maxima, however, run closer to 72 days before inferior conjunction. This one, as estimated by interpolation from data in the Tuckerman tables, was 73 days before, with 46.99 degrees elongation. The difference of about half of a degree is of the magnitude that I have supposed was noticeable to the Maya observers and was sufficient as an indication to them that the planet was embarking on its return to the sun. Alternatively one might grant them less observational acuity, and suppose that this was within a vague range of positions, any of which might have been judged simply as maximum elongation, admitting to the category any date that fell within say a 13-day range one side or the other of the precise maximum. But the other available candidates for the category, and the intervals between them, give grounds for rejecting a 'vague maximum' hypothesis of this sort and for entertaining seriously the more precise 'departure' hypothesis. These are dates 11 and 12, to which attention is now turned.

The latest of the three Bonampak stelae is the one known as Stela 2, which records two dates and the corresponding two events. The latter of these was an autosacrificial act, wherein the ruler offers blood to a deity or an ancestor. The preparations are depicted in the sculpture, in which the ruler's mother offers him the stingray-spine perforating instrument and the shallow bowl of folded paper which was the standard receptacle for blood drawn from one of one's bodily members, while his wife, a lady from Yaxchilan, holds another such bowl in reserve. Though it is apparent from the inscription that the occasion was an anniversary of something, it is not clear what that was. That it may have been something specifically related but prior to the ruler's accession is suggested by the fact that the other date on this stela, with which the inscription opens, was that accession date. This is date 11 of Table 2. As can be seen from the table, it was 14 days after the date of precise maximum elongation of the evening star, as best this can be judged; and during this interval the elongation had diminished by about three-quarters

of a degree. Like date 13, this one also could be a first perceived move
away from the extreme. Interestingly, the interval between this accession
date on Stela 2 (date 11) and the necessary reconstruction of the second
date of Stela 3 (date 13) is 9.13.4, or 3504 days, which is an integral
multiple of 584. Since they concern the same ruler, it is difficult not
to suspect that this was intentional. One may even imagine what might
have been in the obliterated event phrase; a reference, perhaps, to the
sixth Venus anniversary of the ruler's accession. The bound captive may
have been intended to have a part in that commemoration.

Date 12 of Table 2 is from another site, Piedras Negras in the
western Peten, on the Guatemalan side of the Usumacinta River. It is one
of the dates recorded on Throne 1 of that site, and it has the 'star-over-
shell' glyph (as in Fig. 2 b) in the associated event phrase. This suggests a Venus phenomenon. As can be seen in the table, it is another
obvious member of the 'departure from maximum eastern elongation' set.
And it is exactly halfway between the two dates from Bonampak which have
just been considered. From date 11 to date 12 is an interval of 3 x 584
days, and from date 12 to date 13 is another 3 x 584 days.

In slight, but I think significant, contrast to these of the
'departure' category are those that, for the sake of a concise label, I
have called the 'arrival' category (though 'resting' or 'waiting' might
have been more accurate as a description). So far, they are only two:
date 9, from Palenque, and date 10, from Bonampak. Date 9 (like date 2 of
Table 1) is a katun ending from the middle panel of the Palenque Temple of
Inscriptions. As in the case of other katun endings, it would have been
recorded anyway. But as mentioned earlier in connection with dates 2 and
3, there are instances of the notation of Venus phenomena in katun-ending
records when these happened to coincide. In the passage which contains
date 9, the notation is quite explicit. It has the Venus-skull glyph
(nominal, as in the date 2 passage from the same inscription--cf. Fig. 1d)
followed by the 'star-over-shell' glyph (verbal), these being followed
then by phrases with first the 'east' and then the 'west' direction glyphs,
and then finally by a compound glyph with a sitting sleeping figure who
has his head buried in his arms, which are clasped around his knees. The
east-west sequence would seem to imply reversal. Reversal there is, not
of the side of the sun, but of the direction of movement: eastward away
from the sun (or toward a horizon landmark for maximum eastern elongation),
and westward toward the sun (or away from the extremity landmark). The

sleeping figure may very well symbolize the wait at the resting place in between; for that is exactly the situation of Venus on the date of that katun ending. At the 1978 Round Table meeting at Palenque, Michael Closs proposed a similar interpretation of this passage (Cf. Closs 1979, 1982); but I was not then of a mind to anticipate this kind of an astronomical record in a Maya inscription, and I remained skeptical. The picture looks somewhat different now.

Date 10, from Bonampak, is recorded on a piece that is known as Sculptured Stone 1. It is an accession monument, as both the glyphic text and the iconography declare. The date, which had been a difficult one to interpret, was successfully deciphered by Peter Mathews (1980). It is another one that has Venus right at the point of maximum eastern elongation. The interval between date 9 and this one is 11.6.8, or 4088 days, which are equal exactly to 7 x 584, or to 7 x 583.92 plus a 0.56 fraction of a day. (In Table 2 this appears as 7v +1, because of the 'nearest whole day' convention for disposing of fractions.

FIRST STATIONARY POINT

With the amount of attention to Venus that is manifest in the dating of undertakings at Bonampak, the lunar interest that it was possible to see in the initial date of the mural text of Room 1 now seems almost an anomaly; and one is led to wonder whether that date too might not have held something of interest in connection with Venus. The date, it will be recalled, was 9.18.0.3.4, 10 Kan 2 Kayab, which by the assumed correlation was December 10 of the year 790. Consulting again the tables, one finds that on this date the planet, as evening star, was exactly at its first stationary point. It was 21 days before inferior conjunction, and the planet's eastward progress across the field of the fixed stars had come to a halt. From here it would reverse itself to begin the retrograde motion that would carry it through inferior conjunction and, as morning star, after yet another 21 or 22 days, on to the second stationary point; whereupon it would stop, change direction again, and resume its direct or orthograde motion. If the selection of this kind of a date for the event that is commemorated in the first part of the mural text was by design rather than accidental, it indicates that the Bonampak Maya were using more than one reference sphere against which to chart the progress of the planet; for the stationary points are in a way analogous to the maximum elongations, with the difference that they are critical points in the

planet's path in relation to the equatorial sphere, with the fixed stars
as landmarks, whereas the maximum elongations are in relation to the
horizontal sphere, with--for primitive astronomy--topographic features
of the horizon or constructed stationary sighting frames furnishing reference points for charting the planet's positions. The ancients of the
Old World observed the stationary points and noted them in records; so
it should not be too surprising to find indications of the same in the
New World.

INFERIOR CONJUNCTION AND HELIACAL RISING

In Table 3 are listed five dates, of which three are from
Bonampak, one from Seibal, and one from Piedras Negras. The dates pertain
to two categories: (A) inferior conjunctions, and (B) probable heliacal
risings of the morning star. The columns of the table are as before,
except for the last two, which pertain only to category B. They are:
(f) the western elongation of Venus, and (g) the number of days after
inferior conjunction. The numbering of the dates continues from the last
of the previous table. They are listed in chronological order within each
category separately. The discussion again will follow a different order.

Date 16 is the battle date in the mural of Room 2 at Bonampak,
the one whose reconstruction initiated this inquiry, and which I believe
can now be regarded as secure. Its attributes were discussed earlier in
the paper. It was at an inferior conjunction of Venus, as well as at a
zenith passage of the sun and a conjunction of Venus with Regulus, apparently seen as a propitious date for the undertaking, the outcome of which
was such as to confirm the opinion.

Date 15 is from the hieroglyphic steps at Seibal. It was the
date of a Maya ball game. Such events were also the occasion for sacrifices, and ballgame iconography often shows a human figure inside of a

Table 3.   Inferior Conjunction and Heliacal Rising.

	(a)	(b)	(c)	(d)	(e)	(f)	(g)
A. Inferior conjunction:							
(14)	9.15. 3. 8. 8	36v -4	735 Jan 17	301.07	same	0	0
(15)	9.15.16. 7.17	28v -2	747 Oct 30	220.00	same	0	0
(16)	9.18. 1.15. 5	* *	792 Aug 2	133.62	same	0	0
B. Heliacal rising:							
(17)	9.15. 3. 8.12	29v -2	735 Jan 21	305.12	298.02	7.10	4
(18)	9.17.10. 9. 4	* *	781 May 31	73.14	65.65	8.48	5

large rubber ball as it rolls down the flight of stairs that forms the
side of the ball court, to be struck in a moment by the yoke-girded player
poised below. A hieroglyph based on this stereotypic representation
designates sacrificial ball games of this sort. The event ascribed to
date 15 is denoted by this glyph, with an infixed large-toothed skull
suggestive of a Venus association, and with the king of Seibal as the
apparent victim and his erstwhile captor as officiant. The date coincides with an inferior conjunction of Venus. Compared with date 16, the
Bonampak battle date, it is prior by an interval equal to within two days
of 28 mean Venus periods. The named victim is the same king who was cap-
tured nearly twelve years earlier as a result of the raid initiated on
date 4 (Table 1), the 'star-over-Seibal' date that first led to recognition of the 'appearance of the evening star' category.

Date 18 is from the same Piedras Negras throne as was date 12
(Table 2). It is the next one thereafter, and its relation to the previous one is indicated by a recorded distance number of 3.3, or 63 days.
That date was marked by a 'star-over-shell' glyph, probably representing
a military exploit timed to coincide with the apparent first move of the
evening star away from its position of maximum eastern elongation. Date
18 was the occasion of the accession to the throne by the same man who was
involved in the previous event, and the accession was apparently timed to
coincide with a heliacal rising of Venus as morning star. It was four or
five days after inferior conjunction, and the elongation of the planet
amounted to 7.10 degrees of longitude and about a degree and a half of
latitude. That the protagonist was aware of the Venus position, and that
the timing was deliberate, is implied by the context with its 'star' marking of the preceding date.

Dates 14 and 17, just four days apart, are from Lintels 2 and
1, respectively, at Bonampak. These lintels, like Lintel 3 from the same
place (cf. date 5, Table 1), depict a warrior in the act of administering
a symbolic coup to a captive. All three figures have hanging from their
necks the pendant mosaic jade skull mask with large teeth that has Venus
associations. This fact, together with a glyphic phrase which I had considered to be the beginning of their appellatives, had led me to identify
the three figures as representing one and the same individual; in which
case the recorded calendar-round dates require the chronological place-
ments assigned to them here. Recently, however, Peter Mathews (1980) has
published very cogent reasons for considering that the figures may repre-

sent three different persons, with identifications for those on Lintels 1 and 2 that would require placement of their dates one calendar round later. The question is not yet resolved. It hinges on a hieroglyph whose meaning may or may not be what has been supposed. If the dates turn out to belong where I have them here, then date 14 is exactly right for an inferior conjunction of Venus, and date 17 for a heliacal rising of Venus; and the two lintels would then have the Venus association seemingly implied by the pendant masks and by their similarity to Lintel 3 whose date is secure and is clearly associated with Venus. If, on the other hand, they properly belong a calendar round later, then they would have no significant associations with Venus (they would be about three and seven days after superior conjunction, an unlikely time for astrological attributions). The fact that their erroneous placement produced such seemingly significant results as in dates 14 and 17 would then exemplify no more than the sometimes ironic turns of chance. And then one would wonder how many of the others might be merely tokens of the favors of chance.

CONCLUSION

The foregoing data, even at worst, I believe sufficient to indicate that the Maya had attained some fair precision in their astronomical observations, just as the Dresden Codex gives evidence of their attainments in mathematical reasoning with the data. And I think the cases reviewed here also give an idea of why a people in a 'primitive kingdom' stage of social and political development were interested in astronomy, and why it was fostered by those who held power. Finally, I believe it offers strong support for the Thompson correlation, specifically the value 584285 which he first proposed.

POSTSCRIPT: THE TURTLE AND THE PECCARIES

The 16th-century manuscript of the Motul dictionary (John Carter Brown Library, Providence, R.I.) has the following entries:

    ac: tortuga, galápago, ycotea.
    ac, 1. ac ek: las tres estrellas juntas que están en el signo
        de géminis, las quales con otras hazen forma de tortuga.

These explain the three stars on the back of the Bonampak turtle (Fig. 3). A modern-day informant told Thompson that the turtle constellation is Orion (Thompson 1950: 116). The term 'signo' is used in both Spanish and Maya sources for calendrical zodiacal divisions, not for constellations. By the sixteenth century, when the Motul dictionary was being assembled, the constellation of Gemini, because of precession, was already mostly in the 'sign' of Cancer, and the stars in the 'sign' of Gemini included those of Orian and much of Taurus. The Bonampak turtle, then, must represent Orion; and its three stars are there for an obvious reason. So also is the turtle itself there for a reason; it is appropriate to the 8th-century early August date of the battle depicted below it. The peccaries and the other figures are not yet securely identified. (Note that the interior ridge line of the vaulting cannot represent the ecliptic; for that would not be in accord with the orientation of the building, and it would put the turtle on the wrong side.)

Fig. 1. Hieroglyphs denoting Venus.

(a)   (b)   (c)   (d)

Fig. 2. Hieroglyphs of 'star over earth', 'star over shell', and 'star over Seibal'.

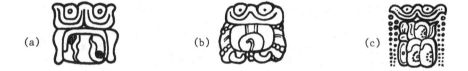

(a)   (b)   (c)

Fig. 3. Cartouches of the northeast vault of Room 2. After Tejeda (1955), modified in accord with copy by R. Lazo in National Museum of Anthropology, Mexico, and with information from M. Miller (1981: 100) based on examination of photographs.

REFERENCES

Closs, M. P. (1979). Venus in the Maya world: glyphs, gods and associated phenomena. In Tercera Mesa Redonda de Palenque, Part 1, ed. M. G. Robertson & D. C. Jeffers, pp. 147-65. Palenque, Chiapas, Mexico: Pre-Columbian Art Research Center.

Closs, M. P. (1982). Venus dates revisited. Archaeoastronomy, 4, no. 4. College Park, Md.: University of Maryland.

Dictionary of Motul. See Martinez Hernandez, J., ed. (1929).

Graham, I. (1967). Archaeological Explorations in El Peten, Guatemala. MARI Publ. 33. New Orleans: Tulane University, Middle American Research Institute.

Kelley, D. H. (1977). Maya astronomical tables and inscriptions. In Native American Astronomy, ed. A. F. Aveni, pp. 57-73. Austin, Tex.: University of Texas Press.

Kelley, D. H., & Kerr, A. (1973). Mayan astronomy and astronomical glyphs. In Mesoamerican Writing Systems, ed. E. P. Benson, pp. 179-215. Washington, D.C.: Dumbarton Oaks.

Lounsbury, F. G. (1978). Maya numeration, computation, and calendrical astronomy. In Dictionary of Scientific Biography, ed. C. C. Gillispie, vol. 15, Supplement I, pp. 759-818. New York: Charles Scribner's Sons.

Martinez Hernandez, J., ed. (1929). Diccionario de Motul, Maya-Español. Merida, Yucatan, Mexico.

Mathews, P. (1980). Notes on the dynastic sequence of Bonampak, Part 1. In Third Palenque Round Table, Part 2, ed. M. G. Robertson, pp. 60-73. Austin, Tex.: University of Texas Press.

Miller, M. (1981). The Murals of Bonampak, Chiapas, Mexico. Doctoral dissertation, Yale University. New Haven.

Robertson, M. G. (1980). The Giles G. Healey 1946 Bonampak photographs. In Third Palenque Round Table, Part 2, ed. M. G. Robertson, pp. 3-44. Austin, Tex.: University of Texas Press.

Ruppert, K., Thompson, J. E. S., & Proskouriakoff, T. (1955). Bonampak, Chiapas, Mexico. CIW Publ. 602. Washington, D.C.: Carnegie Institution of Washington.

Teeple, J. E. (1930). Maya astronomy. Contrib. to Amer. Archaeology, 1, no. 2, 29-116. CIW Publ. 403. Washington, D.C.: Carnegie Institution of Washington.

Tejeda, A. (1955). Ancient Maya Paintings of Bonampak. [Authorship of accompanying text unindicated.] CIW Supplem. Publ. 46. Washington, D.C.: Carnegie Institution of Washington.

Thompson, J. E. S. (1935). Maya chronology: the correlation question. Contrib. to Amer. Archaeology, 14. CIW Publ. 456. Washington, D.C.: Carnegie Institution of Washington.

Thompson, J. E. S. (1950). Maya Hieroglyphic Writing: An Introduction. CIW Publ. 589. Washington, D.C.: Carnegie Institution of Washington. (2nd ed. 1960. Norman, Okla.: University of Oklahoma Press.)

Thompson, J. E. S. (1972). A Commentary on the Dresden Codex. APS Memoirs, vol. 93. Philadelphia: American Philosophical Society.

Thompson, J. E. S. (1974). Maya astronomy. In The Place of Astronomy in the Ancient World, ed. F. R. Hodson. Phil. Trans. R. Soc. Lond., A. 276, pp. 38-98. London: The Royal Society.

Tuckerman, B. (1964). Planetary, Lunar, and Solar Positions, A.D. 2 to A.D. 1649, at Five-day and Ten-day Intervals. APS Memoirs, vol. 59. Philadelphia: American Philosophical Society.

LUNAR MARKINGS ON FAJADA BUTTE, CHACO CANYON, NEW MEXICO

A. Sofaer
Washington, D. C.

R. M. Sinclair
National Science Foundation, Washington, D. C.

L. E. Doggett
U. S. Naval Observatory, Washington, D. C.

Fajada Butte is known to contain a solar marking site, probably constructed by ancient Pueblo Indians, that records the equinoxes and solstices (Sofaer et al. 1979 a). Evidence is now presented that the site was also used to record the 18.6-year cycle of the lunar standstills.

Fajada Butte (Figure 1) rises to a height of 135 m in Chaco Canyon, an arid valley of 13 km in northwest New Mexico, that was the center of a complex society of pre-columbian culture. Near the top of the southern exposure of the butte, three stone slabs, each 2-3 m in height and about

Fig. 1   Fajada Butte from the north. The solar/lunar marking site is on the southeast summit.

1,000 kg in weight, lean against a cliff (Figures 2, 3).
Behind the slabs two spiral petroglyphs are carved on the
vertical cliff face. One spiral of 9 1/2 turns is elliptical
in shape, measuring 34 by 41 cm (Figure 4). To the upper left
of that spiral is a smaller spiral of 2 1/2 turns, measuring
9 by 13 cm (Figure 4).

SOLAR MARKINGS
    Throughout the year near midday the two openings
between the slabs form vertical shafts of sunlight on the
cliff face. The daily paths of these dagger-shaped patterns
on the cliff change with the sun's declination. As the patterns intersect the spirals, the equinoxes and solstices are
uniquely marked.

Fig. 2   The slabs from the south.

Fig. 3 Eastern slab casting a sunrise shadow on the spiral. The inset shows the shadow of sunrise at declination +18°4 (i.e., northern minor standstill).

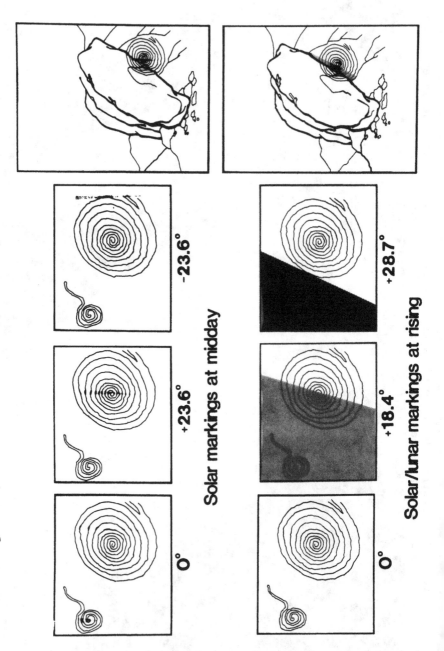

Fig. 4   Solar markings at midday and solar/lunar markings at rising.

At summer solstice one dagger of light descends through the center of the large spiral (Figure 4). On succeeding days the dagger descends discernibly to the right of center. As summer progresses and the sun's declination decreases, the position of the dagger shifts progressively rightward across the large spiral and a second dagger of light appears to the left.

At the autumnal equinox this second dagger bisects the smaller spiral (Figure 4). From fall towards winter the daggers continue their rightward movement, until at winter solstice the two daggers bracket the large spiral, holding it empty of light (Figure 4). Following the winter solstice the cycle reverses itself until the next summer solstice.

In addition to the midday solar markings, we previously noted (Sofaer et al. 1979 b) that at sunrise the inner edge of the eastern slab casts a shadow on the larger spiral (Figure 3). The edge of this shadow crosses the spiral when the sun's declination is positive, and its position shifts leftward an average of 2.5 cm per week between the equinox and solstice (Figure 5). At the equinoxes the shadow edge falls in the far right groove of the spiral. Noting this second possible marking of equinox, we assumed that the sunrise shadows might form a second set of intentional solar markings. However, the locations of the shadows of the summer and winter solstices are of no particular note: at summer solstice the shadow edge falls between the center and the left edge of the spiral (Figure 5); at winter solstice the edge is not on the spiral at all. We noted the location of the shadow edge on the spiral as a sensitive indicator of changing solar declination for half the year, but with no certainty of its use or purpose.

Further observations through the year showed that when the sun is at declination +18°.4 (mid-May and late July), the shadow at sunrise bisects the large spiral, putting the left half in shadow and the right half in light (Figures 4-6). At these times the sun is approximately at the declination of the moon at northern minor standstill. The edge of this shadow is aligned with a pecked groove (Sofaer et al. 1979 a,

Fig. 5   Solar/lunar markings and solar/lunar
risings on the northeast and eastern horizon.

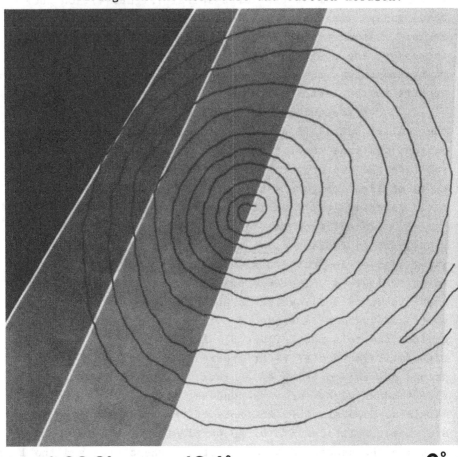

28.7° 23.6°   18.4°   0°

Declinations of solar/lunar
shadow markings

**53.6°   60.3°   67.0°   HORIZON   90°
(East)**

Azimuth of solar/lunar risings

Figure 7) that runs from the spiral's center to the lower left edge, emphasizing this particular occurrence. A further stimulus to search for lunar significance within this assembly was provided by Alfonzo Ortiz (1979), who suggested that because of the dual roles of sun and moon in Pueblo culture, a site in which the sun was so clearly marked would also include the moon.

LUNAR MARKINGS

Just as the rising sun casts a shadow, so too does the rising moon, provided the moon is in the proper portion of its monthly cycle of phases. Thus under the correct conditions (see Construction of the Site) the rising moon at minor standstill casts a shadow bisecting the large spiral.

Fig. 6   Simulated shadow of moonrise at northern minor lunar standstill.

While the sun provided a convenient simulation of
moonrises up to declination +23°4, we had to use artificial
light sources to simulate the moon at higher declinations.
(The next major standstill will not occur until 1987.) We used
a laser for accurate alignment and near-parallel light from a
floodlight to form the shadows. A series of simulations was
calibrated against observations of various sunrises, with corrections made for the obliquity of the ecliptic of AD 1000,
the estimated era of construction of the site. Also taken into
account were the effects of lunar parallax and atmospheric
refraction (Thom 1971) and the height of the eastern and
northeastern horizon.

The simulation of the northern major standstill at
declination +28°7 (epoch AD 1000) showed a shadow falling
tangential to the far left edge of the spiral (Figures 4, 5).

We conclude that the Pueblo Indians recorded the
extreme northern rising positions of the moon at major and
minor standstills. In addition, as we speculated earlier
(Sofaer et al. 1979 a), the number of grooves in the spiral
(counting horizontally from the left edge to the right edge)
may record the length of the cycle. This appears in two possible ways: (1) as the cycle moves from minor to major standstill over 9 to 10 years, the extreme position of the lunar
shadow shifts over the 10 grooves on the left side of the
spiral; (2) the length of the full cycle (18.6 years) may be
recorded by the count of 19 grooves across the full spiral.
The number of grooves may also record a knowledge of the
19-year Metonic cycle. In addition the passage of the shadow
edge through the far right groove of the spiral may record the
midpoint of the declination cycles of the sun and moon.

The following factors can affect the position of the
edge of the shadow for a given sunrise or moonrise (Thom
1971): variations in atmospheric refraction ($\pm$0.4 cm), the
lunar wobble of 9' amplitude and 173-day period ($\pm$0.3 cm),
and variations in lunar parallax ($\pm$0.1 cm). None of these
introduces an appreciable uncertainty in the display.

In addition it is not clear what position of the
rising moon would have been used by the ancient Pueblos to

observe shadows. The difference in the shadow position on the
spiral of two reasonable possibilities - the lower limb just
tangent to the horizon and the lower limb positioned one lunar
diameter above the horizon - is only 0.6 cm. The phase of the
moon does not affect this, as long as a consistent definition
of moonrise is used.

CONSTRUCTION OF THE SITE

We previously presented evidence (Sofaer et al.
1979 a) that the slabs were deliberately placed and were possibly shaped on critical edges to form the midday patterns on
the spiral. In doing so the builders had control over the
placement of the edge that casts shadows at sunrise and moonrise. Calibration of this edge could have required shaping of
the inner surface of the eastern slab (Figure 3). Further
examination of this edge is required to determine whether its
shape is natural or artificial.

Prior knowledge of the standstill cycle would most
likely be necessary to relate the shadow casting edge in
proper orientation to the spiral. This knowledge could most
easily have been gained by accurate horizon watching, a practice reported by McCluskey elsewhere in these Proceedings to
have been extensive among an historic Pueblo group, the Hopi.

Under clear weather conditions a majority of moonrises and moonsets are visible each month, with the shifting
azimuthal position revealing first the moon's monthly cycle
in declination and then, over the years, the standstill cycle.
At Chaco Canyon the monthly declination cycle causes the
azimuth of moonrise to vary from 67° to 113° near a time of
minor standstill. These limits gradually increase over 9.3
years until at major standstill the azimuth varies from 54°
to 126°. On the western horizon the setting cycles mirror the
rising cycles on the eastern horizon.

Near a standstill the azimuthal limits change very
slowly, making it difficult to pinpoint the exact year in
which a standstill occurs. However, this makes possible a
reasonably accurate determination of the amplitude of the
standstill cycle, even if some potential observations are lost

because of bad weather or the moon being at the wrong phase. With a desert environment of clear skies and an unpolluted atmosphere, the Pueblos had the advantage of optimal observing conditions.

Once the standstill cycle was known and the device constructed, the northern limits would be indicated on the spiral by appropriate moonrises that are bright enough to cast shadows. In New Mexico we have observed lunar shadows at moonrise, within a few minutes after the lower limb is tangent to the horizon, from the night after full moon to a few nights after third quarter. Since the moon's phase cycle (29.5 days) is longer than its monthly cycle in declination (27.3 days), the phase at successive declination maxima slowly changes. Thus in any given year there are approximately four shadow casting moonrises occurring near the northern limit for that year (Figure 7).

Fig. 7   Azimuths of moonrises occurring after the end of evening civil twilight and before the beginning of morning civil twilight during the year of a minor standstill. Moonrises with declination within 1° of the standstill limit and with sufficient brightness to cast shadows are circled.

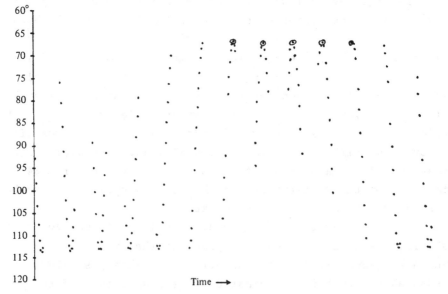

CULTURAL BACKGROUND

The cultural and technological sophistication of the ancient Pueblo Indians of Chaco is evident in their development of an extensive trade and road network and in their planning and building of elaborate multi-story pueblos (Hayes et al. 1981). Interest in astronomical orientation is found in the reported (Williamson et al. 1975) solar and cardinal alignment of several pueblos and kivas (the Pueblo ceremonial structures in Chaco Canyon). It is also interesting to note the possibility of cultural contact with the Mesoamerican societies that had studied eclipse cycles (Lounsbury 1978) and developed complex calendric systems.

In the absence of direct knowledge of the customs of the prehistoric Pueblos, we turn to the historic Pueblos for insights into the ceremonial importance given to bringing together the cycles of the sun and moon. Many ethnographic reports of the scheduling of the winter solstice ceremony indicate strong desire to have the date coincide with the full moon (Stevenson 1904; Bunzel 1932; Ellis 1975). McCluskey (1977) reported that the Hopi synchronized the lunar and solar cycles over 2 to 3 years in setting their ceremonial calendar. More recently McCluskey (1981) has suggested that the Hopis' attention to the moon must have brought them close to observing the standstill cycle: "It would have been a short step for them to look for the moon's house, the theoretical lunistice which the moon reaches every 18.6 years . . ."

Spier (1955) reports that common to most of the historic Pueblos is the starting of the new year with the new lunation closest to winter solstice. Frequent planting of prayer flags at full moon, especially at winter solstice, also indicates the moon's significance in the Pueblos' ritual life (Bunzel 1932). The duality theme in Pueblo cosmology links sun and moon as male/female: sun-father and moon-consort or sister (Stevenson 1904). Ortiz (1981) reports the Tewa Pueblo group as seeing the moon as the mask of the sun.

There was thus a consistent effort to seek the synchronization of lunar and solar cycles. We speculate that there was success in this quest on Fajada Butte in bringing

together at the spiral's center and outer boundary the highest
and lowest positions of sun and moon (Figure 4).

With the possible exception of Casa Grande (Evans &
Hillman 1981), we know of no other evidence of markings of the
lunar standstill cycle in the Americas. We have searched for
other possible explanations for the timing of the shadow phenomena in the culture and weather patterns in Chaco. A scholar
of Chaco's agrarian prehistory (Truell 1981) found nothing
significant in the dates when the sun reaches declination
+18°4, and weather patterns of Chaco do not indicate these
dates as consistent times of rain or other climatic events.
And of course the marking of declination +28°7 is not relevant
to the annual solar calendar. The evidence does point to this
site as a place where ancient Pueblos integrated on one set of
spirals with one set of slabs the cycles of the sun and moon.

ACKNOWLEDGEMENTS

        We are indebted to A. Schawlow and T. Hansch for
supplying the laser used in the simulations; to A. Aveni,
J. Carlson, M. Cohalan, J. Eddy, F. Eggan, G. Hawkins and
J. Young for many fruitful discussions; to K. Kernberger for
photography; to J. McGrath and P. Wier for field work; to
B. Jones, R. Nugent, T. Plimpton and R. White for moonrise
shadow observations. Finally we wish to express our appreciation to W. Herriman, Superintendent, Chaco Canyon, for his
essential help to this research.

REFERENCES

Bunzel, R. L. (1932). Introduction to Zuni ceremonialism. In
    47th Annual Report. Bureau of American Ethnology,
    pp. 467-1086. Washington: Smithsonian Institution.
Ellis, F. H. (1975). A Thousand Years of the Pueblo Sun-Moon-
    Star Calendar. In Archaeoastronomy in Pre-Columbian
    America, ed. A. F. Aveni, pp. 59-88. Austin: Univer-
    sity of Texas Press.
Evans, J. H. & Hillman, H. (1981). Documentation of Some Lunar
    and Solar Events at Casa Grande. In Archaeoastronomy
    in the Americas, ed. R. A. Williamson. Los Altos,
    California: Ballena Press.
Hayes, A. C., Brugge, D. M. & Judge, W. J. (1981). Archaeolog-
    ical Surveys of Chaco Canyon. Washington: National
    Park Service.
Lounsbury, F. G. (1978). Maya Numeration, Computation, and
    Calendric Astronomy. In Dictionary of Scientific
    Biography. ed. C. C. Gillispie, Vol. XV, pp. 759-
    818. New York: C. Scribner's Sons.
McCluskey, S. C. (1977). The Astronomy of the Hopi Indians.
    J. for the History of Astronomy, 8, 174-195.
McCluskey, S. C. (1981). Comment on the paper Stone Age
    Science in Britain? by A. Ellegard. Current Anthro-
    pology, 22, no. 2, 119.
Ortiz, A. (1979). Private communication.
Ortiz, A. (1981). Private communication.
Sofaer, A., Zinser, V. & Sinclair, R. M. (1979 a). A Unique
    Solar Marking Construct. Science, 206, 283-291.
Sofaer, A., Zinser, V. & Sinclair, R. M. (1979 b). A Unique
    Solar Marking Construct of the Ancient Pueblo
    Indians. American Indian Rock Art, 5, 115-125.
Spier, L. (1955). Mohave Culture Items, pp. 16-33. Flagstaff:
    Northern Arizona Society of Science and Art.
Stevenson, M. C. (1904). The Zuni Indians. 23rd Annual Report.
    Bureau of American Ethnology, pp. 9-157. Washington:
    Smithsonian Institution.
Thom, A. (1971). Megalithic Lunar Observatories. Oxford:
    Clarendon Press.
Truell, M. (1981). Private communication.
Williamson, R. A., Fisher, H. J. & O'Flynn, D. (1975). The
    Astronomical Record in Chaco Canyon, New Mexico.
    In Archaeoastronomy in Pre-Columbian America, ed.
    A. F. Aveni, pp. 33-43. Austin: University of Texas
    Press.

THE SKIDI PAWNEE EARTH LODGE AS AN OBSERVATORY

V. D. Chamberlain
Smithsonian Institution, Washington, D. C. 20560

This paper concerns one of four bands of the Pawnee Indians, the Skidi Band, whose history can be traced back to perhaps 1550 A.D. along the Platte River and its tributaries, at about Latitude 41° North, in what is now east central Nebraska in the United States. We are fortunate to have an extensive collection of information regarding the cosmic concepts of the Skidi, most of which was gathered near the beginning of the Twentieth Century. Part of the information has been published and part of it is in archival collections. The most significant single source was just published by the Smithsonian Institution Press (Murie 1981).

The Pawnee lived in earth lodges (Figures 1 and 2) which were made from sod supported by large main posts and smaller poles. The lodges were filled with astronomical symbolism which constantly reminded the Skidi of their beliefs. Indeed, the Skidi lodge was a microcosm which modeled the Skidi concept of the universe outside. The domed roof represented the sky, the dwelling of the supreme god Tirawahat. The floor represented the earth. The four main posts, which supported the roof, symbolized four stars believed to be the pillars of heaven located in the intercardinal directions. These very important mythological stars were associated with certain colors, seasons, times of life and other items (Figure 3).

In the center of the lodge was the fireplace which symbolized the Sun. The lodge was entered through the east oriented tunnel entrance and opposite this, to the west, was the sacred altar, the place of the female Bright Star, also known as Evening Star, believed to be the mother of the first human child and the source of life in general. The west side of the lodge was associated with her fertile garden where it was believed life had originated and where it was renewed each spring. On the altar was a buffalo skull and it was said that the first rays of the rising Morning Star (father of the first human) and the Sun would enter the entryway, pass over the fire symbolically lighting it, and fall on the buffalo skull. Directly over the fireplace was the opening to allow the smoke to exit the lodge; smoke was considered to carry messages from earth to the sky gods. The smokehole was symbolic of a constellation known as the Council of Chiefs (Corona Borealis), a circlet of stars showing the people how their chiefs should sit together to make weighty decisions. Thus the lodge was a place to live in and a place to be aware of the influences believed to have created life, placed it on earth, and to sustain it by the renewing powers from above in the form of light, heat and moisture. Details of the Skidi astrotheology have been reviewed by the author elsewhere (Chamberlain, in press).

The Funk and Wagnall Standard College Dictionary defines the word observatory as, "a building or station for the systematic observation of natural phenomena; especially, one for astronomical observations." Ethnographic evidence suggests that the Pawnee home was a place where systematic observations of the sky were made. James R. Murie (1981:41), the man who gathered most of the information we have about the Skidi, wrote:

> We have no full data as to the precise methods
> of observation, that being part of the profes-
> sional knowledge of the priests; but it is
> said that the usual method was to note the
> positions of certain stars at their first
> appearance after sunset and again the time of
> year when they could be first seen upon the
> horizon at dawn. Observations were also taken
> through the smoke hole of a lodge, by taking
> a seat west of the fire at sunset and noting
> what stars could be seen.

Gene Weltfish (1977:79), in her book The Lost Universe, also wrote:

> The earth lodge served as an astronomical
> observatory and as the priests sat inside
> at the west, they could observe the stars
> in certain positions through the smokehole
> and through the long east-oriented entrance-
> way. They also kept careful watch of the
> horizon right after sunset and just before

dawn to note the order and position of the
stars.

Wedel (1979:86–88) has given archeological evidence relating
the Skidi star symbolism to the Lower Loup phase of Pawnee habitation
which extends from 1550 to 1750 A.D. He noted that 80 percent of Lower
Loup houses had the four center posts and that 88 percent of them faced
due east or very nearly so and that altar platforms were typically found
opposite the entrance. Later Pawnee lodges do not reveal such striking
support for the astrotheology, but the concept of east orientation
persisted. With this east-west orientation in mind, let us see how well
suited the Pawnee lodge was as a naked-eye observatory.

Figure 4 is a geometrical model of a lodge with mostly typical
dimensions: the entryway parameters are not typical, but were selected
to model a lodge where morning sunlight could penetrate into some portion
of the lodge throughout the year; most lodges had longer entryways.

The following are the most important features of the model
lodge: the altar is at point A; the firepit is at C; the center of the
smokehole is at A; a mound of dirt taken from the fireplace when it was
dug is at M directly east of the lodge. The cardinal and intercardinal
directions are marked. The geometry is straight forward and the model,
shown in Table I, was calculated by simple trigonometry.

The Angle α, which defines the limitations of light of the
rising Sun falling directly on the buffalo skull on the altar, is 7.3°
for our model. Skidi mythology indicated that the rays of the morning
Sun would light the skull each morning. It is obvious that this could
not happen. Calculations for our model at latitude 41° north show that
morning light could strike the skull for about 23 days during each
equinox season. Since our entrance tunnel is shorter than typical, this
time period would be even shorter for most actual Pawnee lodges. Sunlight, for our model, could reach the skull starting about mid-March and
continue each clear morning until about one week into April. A few days
into September morning light could again fall upon the altar and this
could continue until almost the end of September. So we see that the
lodge could have served as one method to know the times for renewal of
life in the Spring and for harvest by watching light and shadow through
the entryway.

In this paper we will not consider the interesting play of
direct sunlight through the smokehole and how the Indians might have

TABLE I

Viewing parameters for a model Skidi Pawnee earth lodge with the
following dimensions: diameter = 40 ft.; floor to smokehole = 15 ft.;
smokehole diameter = 2.5 ft.; entryway length = 10 ft., height = 6 ft.,
width = 6.25 ft. Values in the first column are in feet and in columns
2-10 in degrees, rounded to the nearest half degree (see Figure 4).

Oz'	τ	ψ	θ	ω	θ'	ω'	$\bar{\theta}$	$\bar{\omega}$	Δ
0	12.0°	7.5°	84.5°	5.5°	−84.5°	−5.5°	90.0°	0.0°	11.0°
2	11.0	7.0	76.0	14.0	86.5	3.5	81.5	8.5	10.5
4	10.5	6.5	68.0	22.0	78.0	12.0	73.0	17.0	10.0
6	10.0	6.5	61.0	29.0	70.0	20.0	65.5	24.5	9.0
8	9.5	6.0	54.5	35.5	62.5	27.5	58.5	31.5	8.0
10	9.0	5.5	49.0	41.0	56.0	34.0	52.5	37.5	7.0
12	8.5	5.5	44.5	45.5	50.5	39.5	47.5	42.5	6.0
14	8.0	5.0	40.5	49.5	45.5	44.5	43.0	47.0	5.0
16	8.0	5.0	37.0	53.0	41.5	48.5	39.0	51.0	4.5
18	7.5	5.0	34.0	56.0	38.0	52.0	36.0	54.0	4.0
19	7.0	4.5	32.5	57.5	36.0	54.0	34.5	55.5	3.5

noted that in combination with other observations. It would have been
possible to determine an entire detailed calendar by this method and
combining knowledge of direct solar illumination through both the smokehole and entrance with observations made by human eyes leads to truly
impressive possibilities. Discussion of these details are included in
another publication (Chamberlain, in press).

Point 0 is the location of an observer seated on the floor of
the lodge with eye-level 2 feet above the floor. The observer's distance
from the center of the lodge is listed in Table I as OZ'. τ is the
horizontal angle and ψ the vertical angle subtended by the entryway from
point 0. We see from Table I that the observer has a fairly restricted
observational possibility when seated near the center of the lodge and
even more restricted back against the west wall. He could, of course,
move in between the center and the east opening for a less restricted
view and he could sit anywhere within the cone shown in Figure 2 and see
out the entrance. Observations along the east-west axis of the lodge
seem particularly important.

At certain times an observer could watch the rising Sun, the
Moon and planets through the entrance. He could also see a limited
number of bright stars and special features such as the belt of Orion and
the Pleiades, depending on where he sat, and these could have been used
for calendrical purposes.

Columns four through ten of Table I reveal the possibilities
for smokehole observations: θ and θ' are the limiting angles of
elevation and ω and ω' the limiting zenith distances; $\bar{\theta}$ and $\bar{\omega}$ are the
mean angles of elevation and zenith distances for various positions of
the observer; Δ is the angle subtended by the smokehole. We see that its
apparent size shrinks from 11° at the center, in the fireplace, to only
3.5 degrees at the wall.

In order to illustrate the observational possibilities using
the smokehole, we will look at three specific examples: the Seven Stars
(Pleiades), the Chief's Council (Corona Borealis) and the Chief Star
(Polaris). Concerning the beginning of the ceremonial cycle of the Skidi,
Murie (1981:53) wrote:

> The time for the ceremonies of the Evening Star
> bundle was primarily determined by the recurrence
> of the thunder in the spring...it was not the
> very first sound of the thunder...for it might
> have thundered at any time. The approximate
> time was fixed by the appearance of two small
> twinkling stars (the Swimming Ducks)...When
> these could be seen it was time to listen for
> the thunder...Also, the Pleiades began to take
> a certain known position at this time.

Murie (1981:76) further stated that this occurred in Nebraska in early
February. He mentioned the Pleiades again in describing preparations
for planting:

> The priests noted the position of the seven
> stars (Pleiades). At a certain hour of the
> night or dawn, the exact procedure is not
> known to us, a priest sat by the fireside
> and looked up through the smoke hole. If
> he could see the seven stars directly above,
> it was time for the planting ceremonies.

Using our model lodge, it is rather easy to determine the likely procedure the Skidi used.

At 41° north latitude the Pleiades would first become
visible centered in the smokehole seen by an observer seated at the wall
along a radius of the lodge just 3° south of west. This compact little
group of stars would just nicely fit in the 3.5° opening. They would be
due east a short time later at an altitude of 38.3°. Checking the table,
we see that $\bar{\theta}$ would have this value for an observer sitting just a little
more than 16 feet from the center of the lodge. He would, of course, be
due west of the fireplace in the most important symbolic region of the
lodge. Since the angle Δ is only about 4°, the Pleiades would not remain
long in the opening as viewed from this location; they would be gone
within a few minutes. The Seven Stars could be seen from this location
sometime during the night from late July until about the end of December.

The Pleiades would cross the meridian 73° above the horizon.
The table shows that they would be centered in the 10° window for an
observer 4 feet north of the center of the fireplace. The stars reach
this position one hour before sunrise about 10 days into September, and
one hour after sunset about the beginning of February when the Stars known
as the Swimming Ducks first appeared and the annual ceremonies began.

At an azimuth of 225°, the Pleiades would have an altitude
of 68° which would put them in the smokehole viewed from about 6 feet
northeast of the firepit. The observer's back would be toward the northeast pillar of the lodge, the one associated with thunder (see Figure 3).
The stars would reach this position in the sky at the end of evening
twilight in the latter half of February. Perhaps this semicardinal
situation is the one Murie referred to in preparation for planting.

Finally, the Pleiades would be due west at 38.3° elevation
where they could be viewed from about 16 feet east of the center of the
fireplace. They could be seen in this position from late July, one hour
before sunrise, until the end of March, one hour after sunset.

The stars of Corona Borealis, the Chiefs in Council, are very
nearly opposite the Pleiades in right ascension. The Skidi associated
both groups of stars with the zenith. The Chiefs culminate only 13° below
the zenith where our smokehole opening would be about 10°. The group of
stars is about 8° across and would fit nicely into the opening seen from
the very north edge of the fireplace. We see that it was very reasonable
for the Skidi to associate these stars with the top of the sky and with
the smokehole, the group being round like the smokehole. These stars

would occupy this position one hour before sunrise in early February, *the same time of year when the Seven Stars (Pleiades) could be seen from almost the same place in the lodge but at the end of evening twilight.*

Finally, let us consider the star which the Skidi called the Chief Star, the Star-which-does-not-walk-around and which was said to watch over the celestial band of stars to be sure none lost its way and to watch over people on the world below, setting the example of stability for earthly chiefs to follow. The viewing position for Polaris, at its altitude of 41°, would be 15 feet south of the center of the lodge. One of Murie's diagrams of the floor plan showing the order of seating and bundle placement in a lodge during ceremonies has the North Star bundle to the south of the fireplace (Murie 1981:73, Figure 14). At first this might seem inconsistent, but actually the south side is the most likely place for the shrine of the North Star, where the supposed power of the star could literally shine on its sacred representative. It is, of course, the only spot in the lodge where any of the star gods could constantly "look down" through the smokehole upon its shrine or representative.

There are many additional interesting observations to consider, but this is enough to indicate how Skidi priests might have used their houses in their work. We note that similar observational techniques have been known among the Southwest Indians, and we wonder how common they might have been among other Plains Indians as well as elsewhere in Native America and other parts of the world.

Fig. 1   View of an earth lodge in a Pawnee village on the
Loup Fork, Nebraska. (Photo by William H. Jackson,
1871. Courtesy Smithsonian Institution, National
Anthropological Archives)

Fig. 2   The back portion of the interior of a full-scale
          Pawnee earth lodge constructed in the Field Museum
          of Natural History, Chicago. The two visible
          support posts are those associated with the
          intercardinal White Southwest star (left) and Yellow
          Northwest star(right). A buffalo skull sits on the
          sacred altar which is visible at the base of the
          wall between the support posts. A bundle hangs over
          the alter. (Photograph by the author)

Fig. 3   Associations of features of the Skidi earth lodge,
stars, directions, villages, colors, animals,
natural phenomena, seasons, periods of life,
vegetation and other items. (An adaptation of a
diagram from Murie (1981:110) and Weltfish
(1977:112)

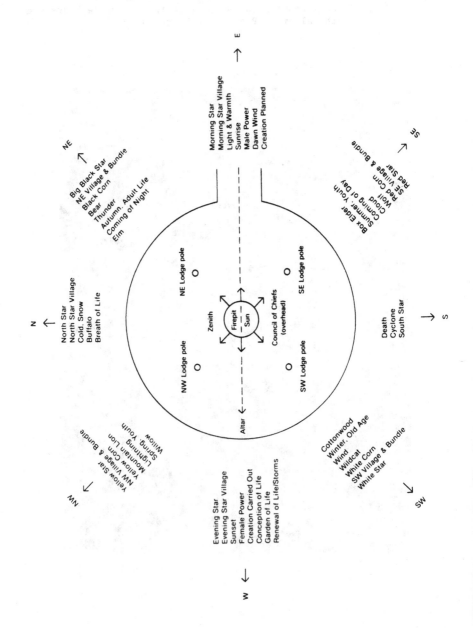

Fig. 4   Astronomical observational geometry of a model
earth lodge.

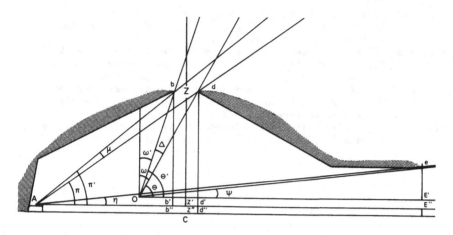

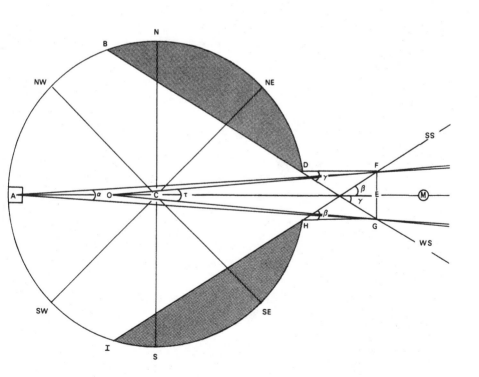

REFERENCES

Chamberlain, Von Del.   In press.   When Stars Came Down to Earth:
     Cosmology of the Skidi Pawnee Indians of North America,
     Ballena Press, expected 1982.
Murie, James R. (1981).   Ceremonies of the Pawnee, Part 1:   The Skiri.
     Edited by Douglas R. Parks, Smithsonian Contributions to
     Anthropology #27, Smithsonian Institution Press.
Wedel, Waldo R. (1979).   House Floors and Native Settlement Populations
     in the Central Plains, Plains Anthropologist: Journal of
     the Plains Conference, 24:85-98.
Weltfish, Gene. (1977).   The Lost Universe, University of Nebraska Press.
     Originally published 1965 by Basic Books, Inc.

CALENDRICAL PETROGLYPHS OF NORTHERN MEXICO

W.B. Murray
Departmento de Antropología, Universidad de Monterrey,
Monterrey, N.L., México

INTRODUCTION

In Mesoamerica the development of astronomical knowledge goes hand in hand with the development of a vigesimal bar-and-dot numbering system which facilitated the recording of long time periods. Indeed, from their earliest appearance in the late Preclassic (Marcus 1976), the Mesoamerican calendrical glyphs are already accompanied by their inseparable companions, the bar-and-dot numbers. The fusion of these two symbol systems, then, must be considered one of the great intellectual milestones of Mesoamerican civilization.

The sophistication of the earliest inscriptions strongly suggests that they are the end product of a long cultural evolution. Yet despite intensive archaeological investigation, no clear antecedents of either calendrical glyphs or bar-and-dot numbers have been identified anywhere in Mesoamerica. The origins of this system remain almost as much a mystery now as when the first clues about its nature were revealed over a century ago. Such information would undoubtedly aid in deciphering many elements of the system which are still poorly understood, but the search for its origins has been so unfruitful that most students of the subject have just about given up hope that anything significant will ever come to light.

Our research attempts to attack this elusive question from a new direction by linking the Mesoamerican glyphs to a different kind of archaeological evidence than has been considered so far. Our working hypothesis is that prehistoric petroglyphs may constitute the missing artefactual link for one of the two elements of the system: the bar-and-dot numbers. Due to the heavy vegetation cover and generally adverse conditions for preservation, such petroglyphs are rarely reported in the Mesoamerican heartland, but in the deserts of the North American Great Basin, they form an important part of the archaeological record. In Nuevo León, on the southeastern edge of the Great Basin only 300 km north of the Mesoamerican frontier, several large rock art sites have been located which provide important new evidence of astro-numerical concepts shared by the two cultural areas.

THE NORTH MEXICAN PETROGLYPHIC COUNTS

Two petroglyphic motifs are especially relevant to this discussion, since they employ separately the two symbols which are later combined to form the Classical bar-and-dot notations. In the rock art literature they are called "tallies, or stroke marks" and "dot grids", and have been recorded throughout much of the Great Basin from southern Oregon right up to the Mesoamerican frontier in the Sierra de Tamaulipas (Swartz 1978; MacNeish 1958). Their numerical function has long been suspected, but proofs to this effect have been hard to come by. After all, any repeated petroglyphic motif can be counted, but we have no way of proving that this numerical property is intentional until we know what was being counted, and can identify systemic rules at work. Moreover, the small quantities represented (usually under 30) and the abstract nature of the symbols used allows for many logical non-numerical explanations. No less an authority than Robert Heizer (Heizer and Hester 1978) interpreted several examples of tallies as representations of game fences, and this explanation may be quite correct in the instances he mentions. Nevertheless, the discovery of a quite unusual tally petroglyph at Presa de La Mula, N.L. (figure 1) raises doubts about the adequacy of this explanation in all cases.

This petroglyph registers 207 tallies on a complex 24-cell grid formed by six horizontal lines and four vertical sections. Its visual impact and complex composition hardly suggests a simple representation of fence posts, nor does the large number recorded seem consistent with a mere tally of animals taken in the hunt. Moreover, the presence of repeated

Fig. 1   Tally Count Petroglyph at Presa de La Mula, N.L., Mexico

quantities in some cells and identical sums on pairs of lines suggested immediately that the numerical properties of the petroglyph were not accidental. The order was never perfect, however, and our initial analysis (Murray 1979) was necessarily vague about the systemic rules involved, for we really had no idea what was being counted.

The possibility of an astronomical explanation, which we initially discarded because of these irregularities, was raised by Aveni (Personal Communication), who pointed out that the total sum (207) was a very good approximation to seven lunar months, and suggested that the count as a whole might refer to lunar observations. At first this explanation did not seem to fit, since the counting pattern generated by the grid could not be correlated to observable points in the lunar cycle. This correlation became evident only when, after closer study of the petroglyph itself, two other "supplementary" glyphs were perceived as part of the tally: an incised circular "completion" glyph and an arrow-shaped "correction" glyph. When these were incorporated into the counting pattern, every element in the count could be related to lunar periodicities observable to the naked eye (Figure 2).We now believe that lunar observation is the best possible explanation of this extraordinary petroglyph.

If this lunar correlation is accepted, three facets of astronomical observation can be inferred which point toward links with Mesoamerican numerical calendrics. First, three different divisions of the lunar month are employed. The first counts from New Moon to Third Quarter (11 + 11), and then adds the Last Quarter (+ 5) as on the first line; the second divides the lunar cycle in half around Full Moon (15 + 13) as on the second line; and the third registers the lunar cycle by quarters (7 + 8 + 7 + 6) as on the third line. Thus, we can

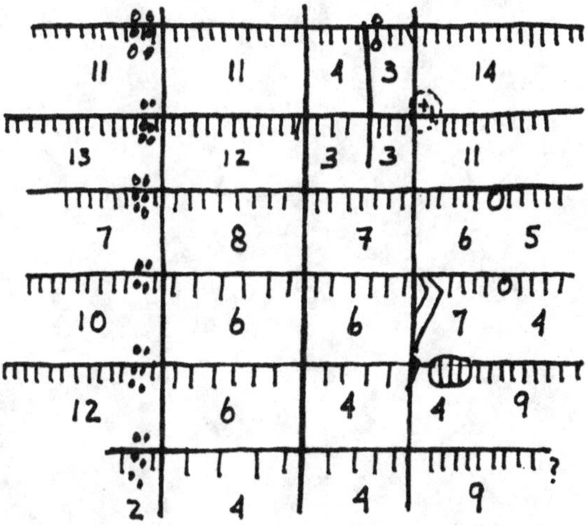

Fig. 2   Numerical Divisions of the La Mula Petroglyphic Count

infer that the total cycle was "built up" from observations of its component parts, and that all observable periodicities were used. Secondly, lunar months are recorded with different total days, ranging from 27 to 30. This variability might be the result of differences in real observations; the objective might have been a precise whole-day estimation of lunar disappearance. But a third feature of the count inclines us to think that these differences may more likely be formulaic and conventional. For if the rest of our interpretation is correct, it can be seen that the tally corrections are introduced to reach the number 148 (the end of the fifth lunar month) and the cluster 176–177–178, the two principal intervals recorded in the Lunar Eclipse Tables of the Dresden Codex (Thompson 1960: 232). This evidence linking the petroglyphic tally to the Mesoamerican context seems unmistakeable, however much it clashes with our present-day archaeological assesment of the North Mexican nomadic cultures.

The evidence of the La Mula tally stone might still be easily dismissed as a statistical accident if no other examples of petroglyphic lunar counting were known, but further explorations in the same general region have revealed at least one other comparable example. At Boca de Potrerillos, another large rock art site some 40 km to the east, a second

Fig. 3   Petroglyphic Dot Counts at Boca de Potrerillos, N.L., Mexico

petroglyph has been located which records the same number 207, this time using the dot motif rather than the tally (Figure 3).We can thus infer that both symbols of the Mesoamerican system were being used by the petroglyphic counters. The internal divisions of this count, however (Figure 4a), are very different from the La Mula tally, and we have no acceptable explanation yet of how they could be generated from lunar observations. In the middle of the petroglyph fifteen large and small dots (11 + 4) might record a lunar half-month, but the three surrounding dot arcs record numbers (63 + 61 + 68) which do not correlate with obvious lunar periodicities. Still, the same total number can hardly be accidental, and greater knowledge of the systemic rules of petroglyphic counting may lead to a plausible solution.

　　　　Nor is the Boca dot count alone. At several other points along the same crest, dot grids register number sums much smaller than 207, some of which could relate to the lunar cycle. One of these is an asymmetrical dot grid (Figure 4b) just to the right of the 207-dot count. The upper part of this grid is regular on the horizontal, and shows an internal division (15+1+11+1) compatible with a type II lunar month centered on the Full Moon, but the five-dot line which follows complicates the issue, and the lower half, which appears to be read on the vertical, shows some frank asymmetries which make any numerical reading order pure speculation. A little down the crest, four other dot grids are found together in a single panel, three of which could refer to the lunar month (Figure 5a–d). Two of them record total

Fig. 4   Schematized representation of Potrerillos Dot Counts

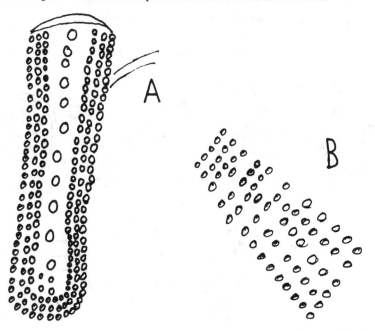

numbers (27 and 29) which are good approximations of the sidereal and synodic month. A third is a half-moon shaped grid which records 42 dots (1½ lunar months), with a possibly significant break (if a serpentine reading order is granted) at the number 29. A fourth dot configuration consists of 18 irregularly spaced dots within a curved loop. This sum does not appear to have lunar significance, but the enclosing line bears a striking resemblance to the shape of Mesoamerican glyphs, and argues for some kind of relationship. On the low rocks of the opposite crest, two other petroglyphs have been identified (Figure 5e) which register possible lunar numbers: 11 + 8 + 11 (+ 1?) = 31, and 11 + 13 (+ 1?) = 25 (poss + 3 = 28). It is interesting to note that both dot grids include the number 11, and if a certain reading order is granted, one of them repeats exactly the 11 + 11 initial month count of the La Mula tally stone. The numerical status of the two large circles and three horizon lines (?) is ambiguous, however, and warns against too-hasty conclusions. If all these symbols are to be summed, a number is reached (59) which would be the closest whole number to two lunar months, a not insignificant achievement in naked-eye observation.

Unfortunately, the issue is not that simple, for many other dot configurations at this site are not symmetrically arranged, and can hardly represent intentional numbers. Olson (in press) has suggested that some of these may be water symbols, and other possible explanations can not be ruled out. A true evolution from diverse representational forms to numerical regularity may be involved. Or perhaps we have simply not mastered enough of the functional context yet to see how its elements fit together, for observations at another part of the site suggest that the petroglyphs are not randomly distributed in the landscape; and may all have been linked at least indirectly to astronomical observations.

Fig. 5   Other possible Astro-numerical petroglyphs at Boca de Potrerillos

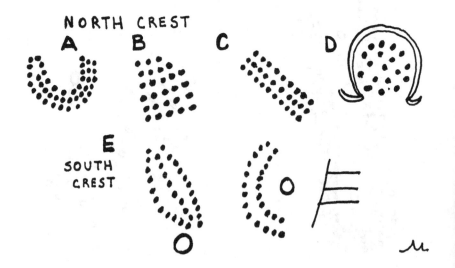

HORIZONTAL ALIGNMENTS

This area is a promontory of rock about 300 m directly behind the canyon mouth which could have served as the base point for various horizon sight lines. Circular motifs of various kinds are especially prominent at this part of the site, and many are deeply carved. From the promontory the eastern horizon presents a striking array of jagged peaks 8–15 km distant which rise well above the true horizon, permitting accurate demarcation of rising points. The nearby crest may have served as an additional foresight. Observations at the March 1980 equinox revealed that the sun rose exactly in the middle of the canyon mouth which cuts the crest, and several petroglyphs were identified which showed possible alignments to this point. Solstice observations the following June did not reveal any clear alignments, but did call attention to two other features of the overall site context. Nearly all the petroglyphs on the (invisible) eastern side of the crest would be comprehended within the angle established by the two solstitial rising points as seen from the promontory. Likewise, every petroglyph at this very large site is carved on the east face of the crest, while equally suitable rock on the west side is totally uncarved. The exact details of the sighting system and its relationship with the carving of petroglyphs still escapes us, but it seems clear that many solar, lunar and stellar rising points (and setting points) could have been noted, and that any of the petroglyphs could conceivably be related (directly or indirectly) to the alignments taken from the promontory.

CONCLUSIONS

Our growing appreciation of the complexity of these petroglyphs leads naturally to a final question. Just who were these petroglyph counters? Are they really the work of nomadic peoples like those found in the region at contact? Or were they the work of itinerant Maya astronomers on country holiday? Are they a pale reflection among savages of the contemporary wonders to the south? Or are they really traces of the archaic roots from which the Mesoamerican system must have developed? The answers to these questions are necessarily uncertain, since petroglyphs are technically undateable, and other archaeological remains have rarely been preserved in close association with the rock art. Nevertheless, a wide range of patinas are encountered at most sites, permitting some rough relative dating, and the motifs themselves provide some surprisingly eloquent clues, all of which seem to point in the same direction.

Taken as a whole, the rock art of northern Mexico is strikingly abstract. At the sites we have discussed clearly representational petroglyphs do not exceed 5% of the total, but one prominent group of representations shows projectile points of various kinds, most of them done in a distinctive incised relief style. (One example is on the La Mula tally stone just to the left of the count.) This same style has been noted at Fort Hancock, Texas in association with Archaic hunting occupations (Sutherland & Steed 1974), and many of the same point types have already been identified archaeologically in north Mexico in Archaic horizons (Nance 1971; Heartfield 1975). Antler shapes and animal paws are also represented with some frequency (Figure 6) and at Boca de Potrerillos an antler headdress is depicted very similar to one found in a burial at Cueva de La Candelaria in the Laguna region of central Coahuila (Aveleyra et al 1956). The spectral face beneath the antler headdress seems to be that of the Archaic hunting shaman. Our best guess now is that he, rather than a Maya astronomer, may have created these petroglyphs, and our question is whether the relationship between the two is one of ancestral paternity, or contemporaneous fraternity.

Much more investigation will be required before we can shed light on the featureless faces of these Archaic men. Many more dot grids and tallies await documentation and analysis at the more than 20 other sites near the ones we have discussed, and in fact much collateral evidence has been omitted here for the sake of brevity. The final analysis of these data will take time, and may well lead to an appreciation of much greater astro-numerical

Fig. 6   Three triangular projectile points with antlers above, and linked
         circles associated

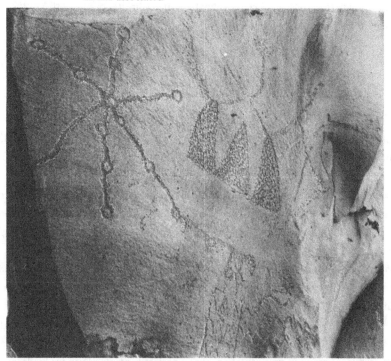

complexity than we are able to perceive today. Most enticingly, however, these studies may lead us closer to the conceptual dawn of Mesoamerican astronomical numbering, and tell us a little more about the role which Archaic shamanism may have played in its gestation.

ACKNOWLEDGEMENTS

The author gratefully acknowledges his debts to Ing. Boney Collins of Monterrey for his initial orientation to Nuevo Leon rock art sites; to Drs Thomas Hester, Anthony Aveni, and Jon Olson for their comments and criticism of our interpretations; to the staff of the Centro Cultural Alfa of Monterrey for field assistance; and to students too numerous to mention for their warm companionship and assistance on numerous field visits.

REFERENCES

Aveleyra Arroyo, Luis; Maldonado, Manuel; and Martinez del Rio, Pablo. (1956). Cueva de La Candelaria. Memoria V. Instituto Nacional de Antropología e Historia. México: Secretaría de Educación Pública.

Heartfield, L. (1975). "Archaeololgical Investigations of Four Sites in South western Coahuila, Mexico" Bulletin, Texas Archaeol. Soc. 46: 127–77.

Heizer, R.F. & Hester, T.R. (1978). "Two Petroglyph Sites in Lincoln Country, Nevada" In Four Rock Art Studies (ed. Clewlow), pp. 1–44. Socorro, N.M.: Ballena Press.

MacNeish, R.S. (1958). Preliminary Archaeological Investigations in the Sierra de Tamaulipas, México. Transactions, Amer. Philos. Soc., 48, pt. 6.

Marcus, Joyce. (1976). "The Origins of Mesoamerican Writing" In Annual Review of Anthropology (ed. Siegal, Reals and Tylor), pp. 35–68. Palo Alto: Annual Reviews Inc.

Murray, W.B. (1979). "Description and Analysis of a Petroglyphic Tally Count Stone at Presa de La Mula, N.L., México" Mexicon 1 (1): 7–9.

Nance, C.R. (1971). The Archaeology of La Calzada: a Stratified Rock Shelter Site, Sierra Madre Oriental, Nuevo León, México. Ph.D. Diss., Department of Anthropology, Univ. of Texas, Austin.

Olson, Jon L. (1981). "Boca de Potrerillos, a Petroglyph Site in Northeast México" Ms.

Sutherland, K. & Steed, P. (1974). "The Fort Hancock Rock Art Site Number One" The Artifact 12 (4).

Swartz, B.K. (1978). Klamath Basin Petroglyphs. Socorro, N.M.: Ballena Press.

Thompson, J.E.S. (1960). Maya Hieroglyphic Writing. Norman: Univ. of Oklahoma Press.

CASA RINCONADA, A TWELFTH CENTURY ANASAZI KIVA

R.A. Williamson
Center for Archaeoastronomy, University of Maryland, College
Park, MD 20742, United States

Sometime near the beginning of the 12th century A.D., the
Anasazi who inhabited Chaco Canyon, New Mexico, built the "great kiva,"
Casa Rinconada. A circular religious building, structurally related to
other great kivas that are found in or near the major villages of the
Canyon, Casa Rinconada stands alone atop a low natural talus ridge that
juts out from the south wall of the canyon. Though separated from other
nearby Pueblo structures physically and stylistically, it stands adjacent
to a densely populated region of the canyon. It is nearly 20 meters in
diameter, 4 meters deep, and is constructed of narrow sandstone slabs
laid up in the rubble core style characteristic of village construction
techniques of the same epoch.

This paper examines the archaeological and structural
evidence for intentional astronomical alignments in Casa Rinconada
(Williamson, et al. 1975; 1977; Fisher 1978) and relates these alignments
to historic Pueblo practice. The totality of the evidence indicates that
Casa Rinconada, built and used during the flowering of Chaco Canyon
culture, was meant to serve as an earthly image of the celestial realm;
that it was not an observatory, but a ritual building whose structure
reflects the central place astronomy had in ancient Pueblo religious
practice.

Throughout the paper I use the term "astronomical alignment"
to denote an observable or measured arrangement of kiva features along a
line of astronomical significance. Such alignments may or may not be
intentional; indeed, the question of whether or not they may have been
placed in the structure intentionally is the subject of this paper.

THE ARCHAEOLOGICAL EVIDENCE

As it stands today, Casa Rinconada is reconstructed from the ruin that was originally excavated in 1931 and 1932 (Vivian & Reiter (1960) and subsequently restabilized at intervals (Richert 1957; Shiner 1959; Mayer 1967). As such, until recently, there was doubt about the

Fig. 1   Casa Rinconada
          (drawn by Snowden Hodges)

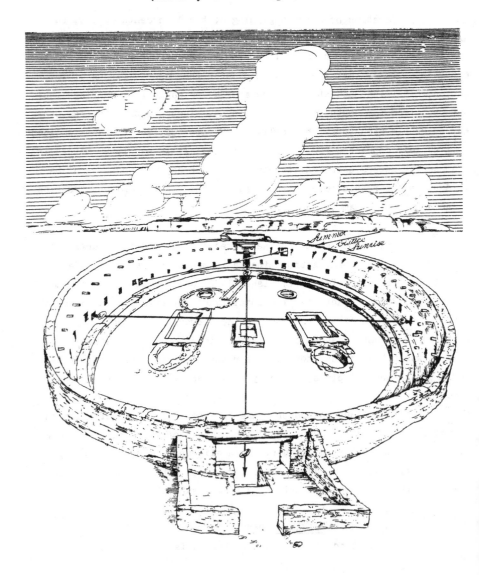

accuracy of the positions of kiva features. Fortunately, though lost for
some time, the original data describing the placement of all the major
interior kiva features as they existed at excavation have been recently
found (Museum of New Mexico n.d.). As will become clear in the following
discussion, they agree for the most part with measurements that my
colleagues and I have gathered in the reconstructed building (Table 1).

Casa Rinconada exhibits a number of structural features (Figure
1) that are similar to those found in other Chaco Canyon great kivas, but
which are not found in the other, smaller kivas (3-5 meters diameter).
The kiva wall contains 28 wall crypts or niches approximately 30 cm wide,
40 cm high, and 30 cm deep, spaced regularly about the kiva about 2 meters
above the dirt floor and one meter above the slab stone "bench" which
surrounds the interior of the kiva. It also contains 6 other slightly
larger niches (A - F), 2 on the east side and 4 on the west side of the
kiva, placed symmetrically about the east-west axis. In addition, there
are two other major wall features, a portal about 75cm on a side in the
northeast, and a niche (40x60cm) high on the southwest wall. Floor
features include two large vaults, a fire pit, and four post sockets which
were apparently used for seating pillars that supported the massive
original roof. In addition, a long subfloor passageway leads from a
secondary staircase beneath the primary staircase and out into the kiva.
This feature is unique to Casa Rinconada. Vivian & Reiter (1965) speculate
that it may have been used to allow persons to enter the kiva unnoticed
by those within. T-shaped doorways to the north and south and
accompanying stairs complete the catalogue of major interior kiva
features. Doorways and staircases are peculiar to great kivas--smaller
prehistoric kivas and historic kivas are entered through roof portals.

In addition to the preceding interior features, the kiva
also had several rooms attached to the north and to the south. One of
the original rooms lay on the northeast side and would have made the
northeast portal, which now appears to be a window, a doorway instead.
Because there is no means to determine the construction dates of any of
the anterior rooms relative to the kiva, it is not known whether they are
contemporaneous with the interior features or not.

Though the detailed function of Casa Rinconada as well as the
other great kivas is unknown (kivas of this size and complexity are not
found in historic Pueblos), it is thought to have served as a general
meeting place for several different clans or religious groups. Because

of its isolated location apart from any of the major villages, Casa
Rinconada may have served as a central ceremonial place.

Table 1 represents the results of two different kiva
mappings. One column ('70s) lists positions taken by my colleagues and
me of the interior features of the kiva. Linear measurements were made
with a steel tape; angular measurements were made with a surveyor's
transit and reduced to local geographical north by sun sightings. We
placed the transit over our best estimation of kiva center, which we
found by stretching a tape between matching pairs of niches and dropping
a plumb from the bisected center. We determined the midpoint of each
niche from the positions of its vertical sides. The other column ('30s)
contains data taken after excavation, but before reconstruction, by
members of Gordon Vivian's field crew. As reference, they used magnetic
north. Their data contain no indication of where the transit was placed
or how its placement was determined. The crew seems to have found the
center of each niche by eye.

In presenting these data together for comparison, I have
rotated and translated the axes of the earlier data to achieve the best
least squares fit to our own recent data. Except where indicated, the
analysis that follows uses our own recent measurements. Although they
may be slightly affected by small errors of reconstruction or
stabilization, they have the advantage of being reduced to geographic
north, and I believe, are overall more reliable than the earlier ones.

GEOMETRY AND ASTRONOMY
The architectural complexity of Casa Rinconada is typical of
many Anasazi structures. With a profusion of different kinds of
architectural features, it is often difficult to decide how to
investigate the potential for astronomical alignments. Simply to draw
lines between all features without regard to their functional or structural
relationship to one another is to invite consideration of many accidental
alignments. Because of this difficulty, one is tempted to start out
looking for certain astronomical lines within the structure in order to
simplify the task. The problem with this approach is that not only may
considerable bias enter in, but one may also miss other more subtle
features of even greater interest to archaeoastronomy. We had first
discovered that the kiva as a whole was oriented along the meridian
(Williamson, et al. 1975, 1977). Is this fortuitous or intentional?

Table 1.

Feature Centers as seen from Kiva Center
(Comparison of recent data ('70s) with
pre-reconstruction data ('30s).

Feature Small Niche	Angular Positions '70s[a]	'30s
1	11°49'	11°49'
2	22°41'	22°49'
3	34°35'	34°36'
4	45°05'	45°03'
5	56°39'	56°27'
6	67°32'	67°41' [b]
7	78°49'	----
8	89°52'	89°51'
9	101°09'	101°08'
10	112°20'	112°23'
11	123°49'	123°49' [b]
12	135°22'	----
13	145°48'	145°49'
14	156°53'	156°53'
15	192°20'	192°23'
16	203°20'	203°20'
17	214°12'	214°16'
18	225°21'	225°20'
19	236°25'	236°25'
20	247°33'	247°36'
21	258°31'	258°38'
22	269°52'	259°49'
23	280°58'	280°41' [b]
24	291°57'	----
25	303°19'	303°19'
26	314°36'	314°34'
27	325°30'	325°34' [b]
27A	N.A.	----
28	347°46'	347°46'

Feature Large Lower Niche	Angular Positions '70s	'30s
A	27°34'	27°34'
B	151°01'	151°28'
C	208°09'	208°07'
D	251°59'	251°59'
E	284°12'	287°11' [c]
F	336°02'	336°11'

Post hole	'70s	'30s
A	44°37'	44°59'
B	134°26'	134°19'
C	223°53'	224°07'
D	314°52'	314°29'

[a] Probable error due to measurement +03'.

[b] Niche in too poor repair to measure.

[c] Possible 3° error in '30s data. Photos show niche intact

Several other obvious alignments seemed to support its intentional
nature, but because of the dangers just mentioned, we measured every
major interior feature of the kiva.

For Casa Rinconada, organizing and interpreting the data are made more
tractable than they might otherwise be by its structural symmetry, which
even casual observation reveals immediately. Under closer scrutiny
(Figure 2), it appears that the placement of nearly all kiva features was
driven by considerations of symmetry. The north and south doorways form
one line of bilateral symmetry (between east and west) and upper niches 8
and 22 form another (between north and south). Further, the kiva was
constructed with particular care to make the centers of major kiva
features coincide. The post sockets form nearly a square (sides 8.28 +/-
.25m), the center of which falls close to the center of the circular kiva
wall (Figure 2). In addition, the centers as defined by the semi-
diameters of the set of 14 niche pairs (with the exception of niches 14
and 28) fall within a circle only 14 cm diameter at the center (about
0.8% of the kiva diameter). Further, large niches A, B, C, F form a
quasi-rectangle, the center of which falls within the same area, and
niches D and E lie symmetrically on either side of the east-west line.
We can use these symmetries and the geometry to simplify our study of the
kiva by asking what astronomical alignments they reveal (Table 2). Each

Table 2.

Angles Between Features[a]

Lines between Large Niches		Lines between post holes	
B-A:	$359^{\circ}24'$	C-D:	$0^{\circ}29'\underline{+}03'$
C-F:	$2^{\circ}06'$	B-A:	$359^{\circ}29'\underline{+}03'$
F-A:	$91^{\circ}48'$	D-A:	$90^{\circ}33'\underline{+}03'$
C-B:	$89^{\circ}35'$	C-G:	$87^{\circ}50'\underline{+}03'$

Line between north & south doorway
(measured along centers of T-doors) : $359^{\circ}56'\underline{+}11'$

Niche 22 - niche 8: $89^{\circ}52'\underline{+}05'$

[a]Probable errors here due to instrumental error ($\underline{+}01'$)
and ability to repeat measurement.

Fig. 2   Casa Rinconada Ground Plan

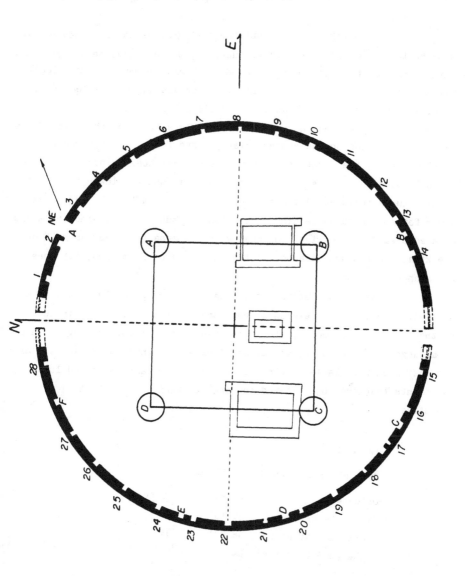

group of features reveals one or more alignments along the cardinal
directions.

The striking fact about these alignments is that they reflect
no visible phenomena: the equinox sunrise could not have been seen from
the interior of the kiva in the days when it was roofed over; nor could
the North-South alignment have pointed to any astronomical appearance,
because in AD 1100 Polaris was about 6o distant from the pole. Only two
astronomical appearances now manifest themselves in or on the kiva (Table
2): 1) at sunrise on the summer solstice, and for at least three weeks
on either side, the light of the sun enters a northeast window, falls on
the wall opposite, and slowly creeps down to illuminate lower niche E for
a few minutes (Figures 1 & 2); 2) for a few days after the vernal equinox
and before the autumnal equinox, just as the sun passes behind the sharp
mesa edge east of the kiva, the shadow of the mesa falls across half of
the kiva. As the sun disappears, the shadow fades from view, but does
not change position.

What other astronomical alignments are possible in the kiva?
What alignments have we excluded from consideration? Given a single
northeast window or doorway, the only sightlines possible through it are
from large niches B, C, D, and E, small niches 13 through 23, and all of
the post sockets. Table 3 presents the most important solar and lunar
alignments that are possible and the architectural orientations I
investigated.

Table 3

Possible Solar and Lunar Alignments

Orientation	Azimuth	Altitude
Niche E - NE portal	$65^{o}15'$	$6^{o}05'$
Niche 22 - NE portal	$58^{o}33'$	$4^{o}34'$
Niche 23 - NE portal	$64^{o}24'$	$4^{o}50'$

Astronomical Object	Azimuths for 0°	3°	6° altitude
Sun (center) (+E)	$59^{o}56'$	$62^{o}39'$	$64^{o}59'$
Moon (center) (+E+I)	$53^{o}57'$	$56^{o}46'$	$59^{o}13'$
(+E-I)	$67^{o}22'$	$70^{o}05'$	$72^{o}19'$

o <u>Solar alignments.</u> The line between niche E and the window
has already been mentioned as fitting a summer solstice
sunrise alignment. Prior to reaching niche E, the solstice
sun also briefly illuminates niche 23 above and south of it.
However, as the light spot passes down the kiva wall, it
falls across niche 23 asymmetrically, whereas it illuminates
large niche E symmetrically before fading. Because of this
asymmetry, and the fact that the placement of niche 23 seems
to be related to the regular spacing between the small
niches, only niche E is left as a potentially intentional
solar alignment. No other solar alignments are possible in
the kiva as it now stands.

o <u>Lunar alignments.</u> Lunar alignments are possible to upper
niches 22, 23, and lower niche E. As the moon reaches the
summer solstice point in its 18.6 year cycle, it would
illuminate niche 23 and niche E, just as the summer solstice
sun does. As it passes further north, it would be capable of
illuminating niche 22 near its northern extremum. On the way
north, its light would also fall about 1 meter above the base
of post D.

o <u>Stellar and planetary alignments.</u> If one were to assume
that the NE portal functioned as the sighting hole for
stellar alignments, features extending from lower niche B to
E would be possible backsights. However, though it is
technically possible to sight stars from some of the niches
through the portal, the large number of features makes the
probability of sighting a bright star or two very high.
Because of this and the total absence of supporting
ethnographic data, I searched for no stellar or planetary
alignments.

ETHNOGRAPHICAL EVIDENCE

Archaeoastronomy in the Americas, and in the Southwest
especially, is aided by the fact that we can still learn from direct
descendants of the peoples whose structures we study. Published
ethnologies and ethnohistories are not only helpful for understanding a
given structure (Williamson 1981a); they are essential if one is to place
the object of study in the context of the culture that produced them.

The ethnological evidence for historic Pueblo interest in astronomy is well documented in the extensive studies from the turn of the century (Fewkes 1892; Stevenson 1904; Cushing 1941), as well as in more recent studies (Ellis 1975; Frazier 1978; McCluskey 1977; Young & Williamson 1981). It divides reasonably well into seven kinds: 1) general knowledge of the motions of the celestial bodies; 2) stories that include astronomical references; 3) ceremonies that have distinct astronomical components; 4) places (shrines) from which the sun and/or moon are watched to determine a calendar; 5) small structures or shrines that are devoted to the sun or moon; 6) whole buildings that are oriented astronomically; and 7) buildings that contain astronomically oriented features for observing the sun or other celestial bodies. Historic Pueblo astronomy is earth centered, related to practical, everyday needs, and indissolubly connected to ritual practice. The motions of the sun are observed most often, followed by the moon, planets, and stars, in that order. Although the stars are not ignored, they are not used for orientation, save for using Polaris as an indicator of North.

Prehistoric Pueblo astronomy seems to have had similar characteristics. Ellis (1975) has argued for a 1000-year tradition of astronomical knowledge among the Pueblos, and considerable archaeoastronomical evidence now exists to support her contention (Reyman 1980; Williamson, et al. 1975, 1977; Williamson 1981b).

A number of facts derived from the historic record are helpful in understanding some of the features of Casa Rinconada. Casa Rinconada is organized symmetrically about the cardinal directions. We know from many sources that the four directions (or six, depending on the Pueblo and the circumstances) are a deep seated part of Pueblo consciousness. Important acts are done in fours: in the ceremonies, the Hopi priest casts corn meal or blows smoke to each of the four directions; the Zuni storyteller repeats important phrases four times; four lines of cornmeal close a path, and so on. However, the four directions of the Hopi and Zuni are the solstitial, not the cardinal, directions. By contrast, for the Pueblos along the Rio Grande, ceremonial directions approximate the cardinal directions. It is not clear whether Western notions have influenced their view of the directions or whether they had developed a notion of the cardinals independent of the west. In any event, both directional systems are evident in Casa Rinconada. The especially strong emphasis on the four

directions in the structure seems to reflect a similar interest on the
part of the Anasazi who built it. The corners of the post sockets, as
seen from the center of the kiva are situated midway between the
cardinals. Though they do not lie precisely along the solstice
directions for the latitude of Casa Rinconada, nevertheless, the
directions of the postholes from the center would be understood by
historic Pueblo Indians as symbolic of the cardinal directions. In their
use of astronomy, precise directionalism generally gives way to
maintaining symmetry.

The summer solstice alignment has its historic counterparts
as well. Kivas and other ceremonial rooms in Hopi (Frazier 1978), Zuni
(Cushing 1941), Isleta (Parsons 1930), and Cochiti (Lange 1958) all contain
portals through which the sun streams at the solstices or at other crucial
times in the calendrical cycle. This appearance is more confirmatory
and/or ceremonial than strictly calendrical, since the sun priest continues
to watch the sun along the horizon to set the calendar. The ritual
importance of the bright light of the sun streaming into a darkened room
is emphasized by the common Pueblo story of a maiden who is impregnated
by sunlight that falls through a portal in her room. The stories vary in
their details, but the outcome is generally a powerful set of twins, sons
of the sun (often referred to as the Morning Star and Evening Star) who
aid the Pueblo people.

DISCUSSION

In the foregoing I have attempted to establish that certain
astronomical alignments do exist in Casa Rinconada, and that they are
consistent with historical practice, i.e., that these alignments would be
familiar to historic Pueblo Indians. The question of whether these
alignments are intentional is as yet unanswered. A statistical argument
is unhelpful in resolving this question because the features of the kiva
are different in kind and cannot be grouped together in a statistical
sample. Yet, because of symmetry, the findings from different features
reinforce each other. Nor are statistics very helpful in comparing this
structure with others like it because Casa Rinconada is not only unique
in the variety of astronomical alignments it manifests, it is also
architecturally unique as well. Of the 3 other excavated great kivas in
Chaco Canyon, only the one in the eastern plaza of Pueblo Bonito is
oriented along the meridian (Williamson, et al. 1977). Though it possesses

certain symmetries, they are not repeated as often as those of Casa
Rinconada, nor are they as clearly organized about the cardinal
directions.   Chetro Ketl and Kin Nahasbas, the remaining excavated
great kivas in Chaco Canyon have no known astronomical alignments.

It is very unclear whether the possible summer solstice or
lunar alignments were intentionally constructed, since the northeast
exterior room may have originally obscured any view of the sun or the
moon. If the room was part of the original structure, the only way in
which the sun or moon could have been seen in the kiva was for there to
have been a second portal through the exterior wall of the room.
Contemporary practices suggest that an exterior portal would not be
unlikely, but there is no evidence to prove or disprove the hypothesis
that the northeast portal was intended for astronomical alignments. If so
intended, it must have been meant to celebrate the arrival of summer
solstice, not to predict it, since the alignment is not a precise one.

Even eliminating the summer solstice or lunar alignments from
consideration, the remaining evidence tells us: 1) the building as a
whole is oriented along the meridian; 2) all of the interior features are
organized symmetrically about the cardinal directions; and, 3) it is
placed west of a mesa edge whose shadow falls across the kiva near the
equinoxes to display the division of the year between summer and winter.
Taken together, these three facts argue that the structure is
intentionally oriented along the cardinal directions.

Further, as a ritual structure, it possesses a certain
uniqueness already which sets it apart from ordinary rooms or groups of
rooms. Historic kivas are often deliberately oriented to the southeast
or southwest, i.e., roughly along the winter solstice sunrise or sunset
directions (Mindeleff 1891). As previously noted, some may also possess
portals aligned to the solstices. Thus it is part of the tradition of
kiva construction that they may be astronomically aligned; if historic
practice reflects prehistoric developments, we would expect to find that
some, but not all, prehistoric kivas were astronomically aligned.

Perhaps the most telling historic evidence for the sort of
alignments we observe in Casa Rinconada is found in part of the creation
myth of Acoma Pueblo, a Keresan-language Pueblo about 150 km south of
Chaco Canyon (Figure 3) whose inhabitants seemed to have derived, at least
in part, from the Chaco area (Ellis 1975). According to this story,
(Stirling 1942) the first kiva was circular in order to represent the

Fig. 3 The Southwest, and
the Pueblo area

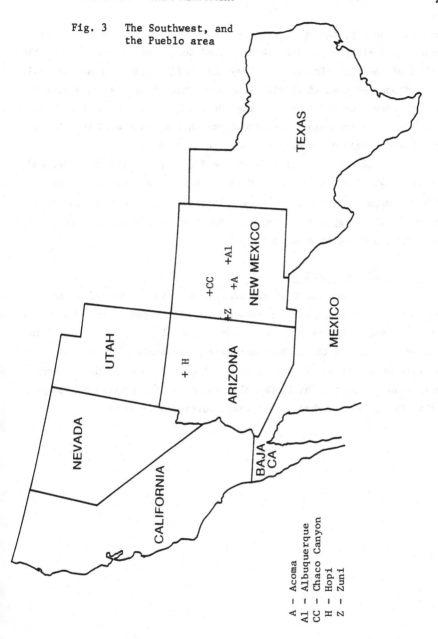

A – Acoma
Al – Albuquerque
CC – Chaco Canyon
H – Hopi
Z – Zuni

circle of the sky; the roof, which was supported by four beams, represented the Milky Way. The makers placed turquoise under the foundation
in each of the four directions. They also built a hollow place or vault
in the floor and covered it with wood upon which to dance. The kiva of
the myth served as a deliberate image of the cosmos. The myth suggests a
tradition of building and orienting kivas that extends well into the past
(the present kivas of Acoma are square, not round).

The resemblances between the "first kiva" and Casa Rinconada
are striking. Taken together with the archaeological findings, this
evidence suggests that Casa Rinconada is conceptually related to the Acoma
"first kiva." It appears to be the product of an Anasazi attempt to image
the cosmos in an earthly building.

ACKNOWLEDGEMENTS

Craig Benson, Clarion Cochran, Marilyn Englander, Howard
Fisher, Frances Goodwin and Jane Young have all helped at one time or
another in measuring the kiva or in understanding its alignments. I am
indebted to them and to the National Geographic Society who made some of
the field work possible through a research grant. I also thank Anthony
Aveni, Jean Monroe and Philip Chandler who read this article and helped
me clarify it for the reader, and Janie Coles who typed it.

REFERENCE LIST

Cushing, F.H. (1941). My Life at Zuni. Santa Fe: Peripatetic Press.
Ellis, F.H. (1975). A Thousand Years of the Pueblo Sun-Moon-Star Calendar. In Archaeoastronomy in
   pre-Columbian America, ed. A.F. Aveni. Austin: University of Texas Press.
Frazier, K. (1978). Solstice Watchers of Chaco. Science News, 114, pp. 148–151.
Fewkes, K. (1892). A Few Summer Ceremonials at the Tusayan Pueblos. Journal of American Ethnology and
   Archaeology, 2, pp. 1–160.
Fisher, H.J. (1978). Geometry and Astronomy at Casa Rinconada. Archaeoastronomy, 1, No. 3, pp. 5–6.
Lange, C.H. (1958). Cochiti, A New Mexico Pueblo Past and Present. Carbondale and Edwardsville:
   Southern Illinois University Press.
Mayer, M.T. (1967). Stabilizations Records, Casa Rinconada, Chaco Canyon National Monument. National
   Park Service Archives, Western Archaeological Center, Tuscon.
McClusky, S.C. (1977). The Astronomy of the Hopi Indians. Journal for the History of Astronomy, 8, pp.
   174–195.
Mindeleff, V. (1891). A Study of Pueblo Architecture, Tusayan (Hopi) and Cibola (Zuni). Bureau of American
   Ethnology Report 8. Washington.
Museum of New Mexico (n.d.) Magnetic Bearings in Casa Rinconada. Miscellaneous Notes from Chaco
   Canyon Field School, 1931–32. Laboratory for Anthropology, Santa Fe.
Parsons, S.C. (1930). Isleta. Bureau of American Ethnology Annual Report, 47, p. 292.
Reyman, J. (1980). The Predictive Dimension of Priestly Power. Tr. Illinois State Acad. Sci. 72, No. 4,
   pp. 40–59.
Richert, R. (1957). Comprehensive Stabilization of Casa Rinconada, a Great Kiva, in Chaco Canyon National
   Monument. National Park Service Archives, Western Archaeological Center, Tuscon.
Shiner, N. (1959). Maintenance Stabilization at Casa Rinconada, Chaco Canyon National Monument.
   National Park Service Archives, Western Archaeological Center, Tuscon.
Stevenson, M.C. (1904). The Zuni Indians. Bureau of American Ethnology Report, 23, pp. 1–634.
   Washington.
Stirling, M.W. (1942). Origin Myth of Acoma. Bureau of American Ethnology Bulletin, 135, pp. 1–123.
   Washington.
Vivian, G. & Reiter, P. (1960). The Great Kivas of Chaco Canyon and their Relationships. School of American
   Research and the Museum of New Mexico Monograph, 22. Santa Fe.
Williamson, R.A. (1981a). Using Ethnology for Archaeoastronomy: a Southwest Example. Presented at the
   International Astronomical Union Archaeoastronomy Symposium, Oxford.
Williamson, R.A. (1981b). North America: A Multiplicity of Astronomies. In Archaeoastronomy in the
   Americas, ed. R.A. Williamson. Los Altos: Ballena Press.
Williamson, R.A., Fisher, H.J., Williamson, A.F., Cochran, C. (1975). The Astronomical Record in Chaco
   Canyon, New Mexico. In Archaeoastronomy in pre-Columbian America, ed. A.F. Aveni,
   pp. 33–42. Austin: University of Texas Press.
Williamson, R.A., Fisher, H.J., O'Flynn, D. (1977). Anasazi Solar Observatories. In Native American
   Astronomy, ed. A.F. Aveni, pp. 203–218. Austin: University of Texas Press.
Young, M.J. & Williamson, R.A. (1981). Ethnoastronomy: the Zuni Case. In Archaeoastronomy in the
   Americas. Los Altos: Ballena Press.